Organic Chemistry
Lab Manual

Organic Chemistry
Lab Manual

(Semi-micro Qualitative Analysis, Preparations, Separations and Estimations)

By

Dr. N. S. GNANAPRAGASAM
Professor and Head of the Dept. of Chemistry (Retd.)
Loyola college (Autonomous)
Chennai - 600 034.

AND

Prof. G. RAMAMURTHY
Vice-Principal & Former Head of the Department of Chemistry
Loyola College (Autonomous)
Chennai - 600 034.

ANANDA BOOK DEPOT

First Published – 1987,1990,1996,1998
Reprint – 2000,2002,2002,2005
Revised Edition – 2006
Reprint – 2007, 2007
Revised Edition – 2008.
Reprint - 2022 2023 - 2026

ISBN : 978 - 81 - 87156 - 19 - 2

New No. 60, Old No. 23, Beerkaran 2nd Street,
Vellala Teynampet, Chennai - 600 086
Phone : 044 - 4851 4646
Mail : mailabd53@gmail.com
Web : www.anandabook.in

Published by : Divya Subramanian for Ananda Book Depot
60/23, Beerkaran Street, Vellala Teynampet, Ch- 600 086

PREFACE

This book entitled "Organic Chemistry : Lab Manual" has been written to fulfil the needs of the B.Sc. (Chemistry) and M.Sc. (Chemistry) students of Indian Universities. In recent years, emphasis has been given to semi-micro methods inorganic experiments, particularly in organic analysis. The semi-micro methods has resulted in a considerable saving of expensive chemicals and laboratory time.

A stencilled edition of organic analysis prepared by us earlier was found to be a very useful guied in the organic lab, both by students and professors. In this printed form, additional preparations, seperations and estimations have been included, to provide a complete guide to organic chemistry practicals.

Suggestions and criticism towards the improvements of this book are most welcome. We sincerely thank our colleagues in Loyola College for thier valuable suggestions and helpful discussions. We are grateful to S.Viswanathan (Printers & Publishers) Pvt., Ltd., for their keen interest in bringing out this revised and totally updated edition.

N. S. GNANAPRAGASAM

G. RAMAMURTHY

CONTENTS

PART A

SEMI - MICRO QUALITATIVE ANALYSIS

INTRODUCTION

In organic qualitative analysis, the group to which the given compound belongs should be identified experimentally. At the end of each analysis, a student should report the following :—

(1) Melting/boiling point of the given substance.

(2) Nature of the substance(s).

  (a) acidic/phenolic/basic/or neutral.

  (b) aliphatic or aromatic.

  (c) saturated or unsaturated.

(3) Elements present (only characteristic elements—N, S and halogens—should be reported).

(4) Functional group/s present.

(5) Derivative.

(6) Melting point of the derivative.

Reactions carried out with the object of detecting the above are called **classification reactions.** They should be carried out first, with known organic compounds of each class, before starting the systematic analysis on unknown samples.

Attention must be paid to the conditions under which each reaction takes place. Organic reactions are very slow reactions. Hence, the student should be patient and thorough in his work.

Organic substances are generally volatile. Volatile liquids (which may readily catch fire) and metallic sodium (which may cause an explosion or a fire, if carelessly thrown into water in the sink) must be handled very carefully. If proper care is not taken in the handling of these substances, there will be the danger of unexpected fires, explosions and bodily injury.

When elements like nitrogen and halogens are found to be absent, the student need not look for the presence of groups containing N or halogens, such as, nitro group, amides, anilides, amines, acid chlorides and halogenated hydrocarbons.

A. Preliminary Reactions

No.	*Experiment*	*Observation*	*Inference*
1.	**Colour**		
	Observe the colour of sub.	Yellow	Aromatic nitro compd. quinone, diketone.
		Dark brown	Aromatic amine, phenol.
		Colourless	Acid, aldehyde, ketone, ester, carbohydrate.
2.	**Smell**		
	Note the smell of sub.	Pleasant or fruity	Ester
		Phenolic	Phenol
		Aniline-like	Aromatic amine.
		Kerosene-like	Hydrocarbon.
		Pungent, irritating	Aromatic compd., with halogen in side chain.
3.	**Litmus test**		
	0.1 g (or 2 dp.) sub. + 2 ml water. Shake well and test soln. with litmus paper.	Blue turns red	Acid or phenol.
		Red turns blue	Amine
4.	**Solubility tests**		
	0.1 g sub. +2 ml satd. $NaHCO_3$ soln.	Dissoves with brisk effervescence of CO_2.	Acid and acid chloride.
5.	0.5 g sub. + 2 ml dil. HCl. Shake well.	Dissolves in HCl, reappears on adding 10 % NaOH dropwise	Basic substance (amine).
6.	0.5 sub. + 2ml 10 % NaOH, Shake well.	Dissoves in NaOH, reprecipitates on adding con. HCl	Aromatic acid, Phenol.
		Dissolves producing a strong yellow colour, which is removed by con. HCl	Nitrophenol, salicylaldehyde.

		Ammonia gas evolved	Ammonium salt, amide.
		Solution turns yellow or brown when boiled	Sugar, aliphatic aldehyde.
		Oily globules obtained having aniline-like smell, when boiled	Anilide.
		Oily layer disappears on boiling: white ppt. obtained on acidification with con. HCl and cooling	Ester and amide of aromatic acid
		The sub. unaffected in cold and on boiling	Nitro compd., ketone, amine, hydrocarbon, aromatic halogen compd.
7.	**Action of con. H_2SO_4** 0.1 g sub. + 1 ml con. H_2SO_4. Warm gently.	Sub becomes black with burnt sugar smell.	Carbohydrate, polyhydric alcohol.
		Violent reaction	Aliphatic or aromatic alcohol.
8.	**Soda lime test** 0.1 g. sub. + 1 g. soda lime. Heat mixture strongly.	Ammonia gas liberated	Ammonium salt or amide.
		Kerosene smell	Acid.
		Phenolic smell	Phenolic acid, ester.
9.	**Neutral $FeCl_3$ test** 0.1 g. sub. + 1 ml water (or alcohol if insoluble in water) +5 dp.* neutral $FeCl_3$. *see Sect. H for prepn.	Violet, blue or green colour	Phenol
		Blue ppt	Naphthyl amine, naphthol.
		Brown ppt	Carboxylic acid
		Strong yellow colour	Hydroxy acid.
10.	**Tollen's reagent** 0.1 g. sub. + 2 ml	Black ppt. or	Aldehyde, reducing

	Tollen's reagent. Heat in water bath	bright silver mirror	sugar, polyhydric phenol or other reducing agent.

B. Tests for Saturation/Unsaturation

11.	0.1 g sub. + 1 ml water +3-4 dp. bromine water.	Yellow colour decolourised	Unsatd. compd.
	Shake well.	Yellow colour persists	Satd. compd.
		Yellow colour decolourised, ppt. formed	Satd. compd. like amine, phenol,
12.	0.1 g. sub. + 3-4 dp. bromine in CCl_4	Yellow colour decolourised	Unsatd. compd.
	(instead of Br_2/H_2O, if sub. does not dissolve in water). Shake well.	Yellow colour persists	Satd. compd.
13.	0.1 g. sub. + 1 ml Na_2CO_3 soln. + 1 % $KMnO_4$ dropwise.	Pink colour decolourised	Unsatd. compd. easily oxidisable compd. (aldehyde)
		Pink colour remains	Satd. compd.

C. Tests for Aliphatic / Aromatic Nature

14.	Ignite 0.1 g. (2 dp). sub. in Ni spatula.	Burns, sooty flame.	Aromatic compd.
		Burns, non–luminous flame	Aliphatic compd.
15.	0.2 g. sub. + a mix. of 1 ml con. HNO_3 and 1 ml con. H_2SO_4. Shake well.	Yellow ppt. or yellow soln.	Aromatic compd.
	Heat 10-15 min. on a boiling water bath. Pour it into 20 ml of cold water.	No yellow colour	Aliphatic compd.

D. Test for Elements

16. Sodium fusion tests

Melt a small piece of dry Na in a small fusion tube. Add 0.1 g. solid (or 3 dp. Liquid) to fusion tube. Heat gently at first, then to red-hotness. Quickly and carefully plunge red-hot end of tube into 10 ml distilled water in China dish. Stir soln. well with broken end of tube, boil and filter. Divide filtrate into 4 parts and test as follows:

(*a*) Test for N : 1 ml fusion ext + few crystals of ferrous sulfate. Boil and cool. Add 2 ml dil. H_2SO_4. Blue/green ppt. or soln. indicates N.

$$Na + C + N \longrightarrow NaCN$$

$$2NaCN + FeSO_4 \longrightarrow Na_2SO_4 + Fe(CN)_2$$

$$Fe(CN)_2 + 4NaCN \longrightarrow \underset{\text{sod. ferrocyanide}}{Na_4[Fe(CN)_6]}$$

$$3Na_4[Fe(CN)_6] + 4Fe^{3+} \longrightarrow \underset{\substack{\text{Ferric ferrocyanide} \\ \text{(Prussian blue)}}}{Fe_4[Fe(CN)_6]_3} + 12Na^+$$

Fe^{3+} in the last step originates from aerial oxidation of Fe^{2+} of $FeSO_4$ soln. (on standing).

Test for S :

(*b*) *Sod. nitroprusside test* : To 1 ml fusion ext, add 1 ml of sod. nitroprusside soln. Violet colour shows S.

$$2Na + S \longrightarrow Na_2S$$

$$Na_2S + \underset{\text{sod. nitroprusside}}{Na_2[Fe(CN)_5(NO)]} \longrightarrow \underset{\text{Violet colour}}{Na_4[Fe(CN)_5(NOS)]}$$

(*c*) *Lead acetate test* : To 1 ml fusion ext, add 1 ml of lead acetate soln. Black ppt. shows S.

$$2Na + S \longrightarrow Na_2S$$

$$Na_2S + (CH_3COO)_2Pb \longrightarrow \underset{\text{black ppt.}}{PbS\downarrow} + 2CH_3COONa$$

(*d*) Test for Halogens : 1 ml fusion ext + 1 ml dil. HNO_3. Boil, then cool. Add 1 ml $AgNO_3$. White ppt. (easily soluble in aq. NH_3) shows Cl. Pale yellow ppt. (partially soluble in aq. NH_3). shows Br. Deep yellow ppt. (insoluble in aq. NH_3) shows I.

$$Na + X \longrightarrow NaX$$

$$NaX + AgNO_3 \longrightarrow AgX\downarrow + NaNO_3$$

$$\underset{\text{white ppt.}}{AgCl} + 2NH_3\,(aq) \longrightarrow \underset{\text{colourless soln.}}{[Ag(NH_3)_2]Cl}$$

Note: The ext. should be boiled well with dil. HNO_3 (especially when N is present) to destroy NaCN, which would interfere by forming insoluble white ppt. of AgCN with $AgNO_3$.

$$NaCN + HNO_3 \longrightarrow NaNO_3 + HCN\uparrow$$

E. Tests for Functional Groups

1. Tests for Acids

(*a*) *Litmus test* : 0.1 g. sub. (3 dp.) + 1 ml water. Shake well. Add a dp. of this soln. to blue litmus paper. If it turns red, an acid is present.

(*b*) $NaHCO_3$ test : To 0.1 g. sub. (3 dp). add 1 ml of satd. $NaHCO_3$ soln. A brisk effervescence with the evolution of CO_2 (which turns lime water milky) shows an acid.

$$\underset{\text{acid}}{RCOOH} + NaHCO_3 \longrightarrow \underset{\text{sod. salt}}{RCOONa} + H_2O + CO_2\uparrow$$

$$\underset{\text{lime water}}{Ca(OH)_2} + CO_2 \longrightarrow \underset{\text{white ppt.}}{CaCO_3\downarrow} + H_2O$$

Note: On passing carbon dioxide into lime water for a long time, the white ppt. formed initially, dissolves(due to formation of calcium bicarbonate).

$$CaCO_3 + CO_2 + H_2O \longrightarrow \underset{\text{colourless soln.}}{Ca(HCO_3)_2}$$

(c) *NaOH test* : 0.1 sub. (3dp.) + 1 ml of 10% NaOH soln. Shake well. All organic acids dissolve when shaken with NaOH. On acidifying with con. HCl, the acid is precipitated, if it is insoluble in water (aromatic acid).

$$\underset{\text{Benzoic acid}}{C_6H_5COOH} + NaOH \longrightarrow \underset{\text{sod. benzoate}}{C_6H_5COONa} + H_2O$$

$$C_6H_5COONa + HCl \longrightarrow C_6H_5COOH\downarrow + NaCl$$

Note: If the given acid is soluble in water (e.g., lower aliphatic acid), precipitation will *Not* take place.]

(*d*) *Esterification* : 0.1 g. (3dp.) sub. + 1 ml rectified spirit + 5 dp. con. H_2SO_4. Shake well and warm gently. Pour the hot mixture into 10 ml dil. Na_2CO_3 soln. Pleasant odour (due to the formation of an ester) indicates an organic acid.

$$RCOOH + C_2H_5OH \xrightarrow{H_2SO_4} \underset{\text{Ester}}{RCOOC_2H_5} + H_2O$$

Note: Sodium carbonate soln. will destroy the excess of sulphuric acid.

(*e*) *Phenolphthalein (HPh) test* : Dissolve 0.1 g. (3dp.) of sub. in 1 ml water, add a drop of HPh, then add dil. NaOH, drop by drop. Pink colour after 2 or 3 dp. shows acid.

(*f*) *Fluorescein test* : Mix 0.1 g. sub. with 0.5 g. resorcinol and 4 dp. con. H_2SO_4 in a hard glass test tube. Heat for 3-5 mts. pour into dil. NaOH. An intense greenish-yellow fluorescence indicates a dicarboxylic acid, which forms an inner anhydride (eg., succinic acid, phthalic acid.)

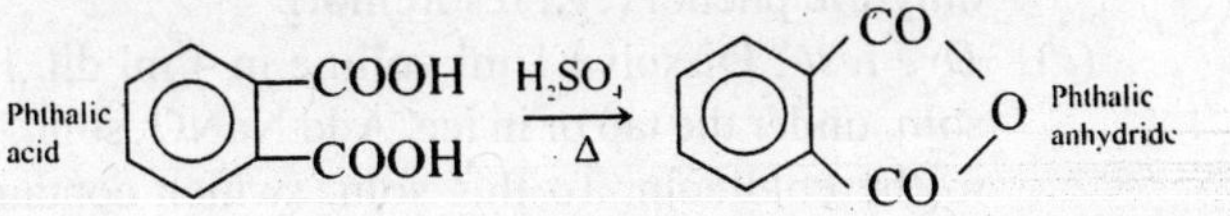

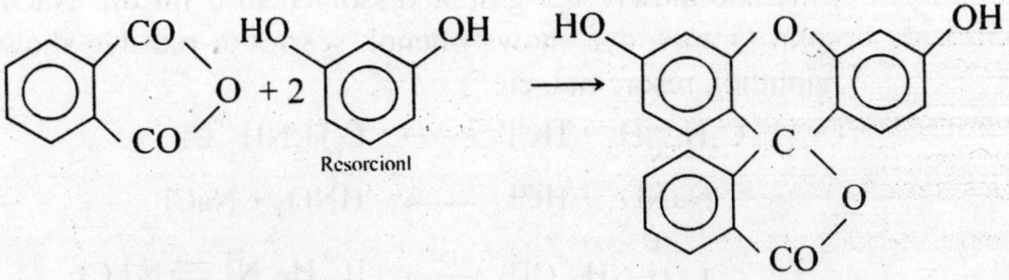

2. **Tests for Phenols :**

(*a*) *Neutral Ferric chloride test* : 0.1 g. (3 dp.) sub. +5 dp. neutral ferric chloride soln. Appearance of blue, violet or green colour indicates a phenol. The colour is due to complex formation between ferric ion and phenol (or substituted phenol).

Note: Prepare neutral ferric chloride soln. as follows : To 1 ml ferric chloride soln., add aq. ammonia, till a permanent brown ppt. of ferric hydroxide appears. Add dil. HOAc, dropwise, till the ppt. just dissolves. If solid ferric chloride is available, dissolve 0.5 g in 50 ml water + 2 dp. HCl. and repeat the procedure described above) (*b*) *NaOH test* : Shake 0.1 g (3 dp.) sub. with 1 ml 10 % NaOH soln. The sub. dissolves and reappears on adding con. HCl, showing the presence of phenol. (This test is common for aromatic acid also.)

$$\underset{\text{Phenol}}{C_6H_5OH} + NaOH \longrightarrow \underset{\text{sod. phenoxide}}{C_6H_5ONa} + H_2O$$

$$C_6H_5ONa + HCl \longrightarrow C_6H_5OH + NaCl$$

Note: Phenols do not give CO_2 with $NaHCO_3$ soln. This test could be used to differentiate phenols from acids.)

(*c*) *Phthalic anhydride test* : 0.1 g (3 dp.) sub. + 0.1 g phthalic anhydride + 3 dp. con. H_2SO_4. Heat gently for 2-3 mts., and pour into 20 ml dil. NaOH soln. Appearance of pink colour (due to phenolphthalein) indicates monohydric phenol ; greenish yellow colour (due to fluorescein) indicates dihydric phenol (eg., resorcinol).

(*d*) *Dye test* : Dissolve 1 ml aniline in 4 ml dil. HCl. Cool the soln. under the tap or in ice. Add $NaNO_2$ soln., drop by drop, to the cold soln. To this soln. (which contains diazonium salt) add slowly 0.5 g sub. dissolved in 5 ml dil. NaOH soln. Orange dye shows phenol. scarlet or red dye shows naphthol, resorcinol, etc.

$$C_6H_5NH_2 + HCl \longrightarrow C_6H_5\overset{+}{N}H_3\,Cl^-$$

$$NaNO_2 + HCl \longrightarrow HNO_2 + NaCl$$

$$C_6H_5NH_3^+Cl^- \xrightarrow{HNO_2} \underset{\text{Benzene diazonium chloride}}{[C_6H_5{-}N^+{\equiv}N]\,Cl^-}$$

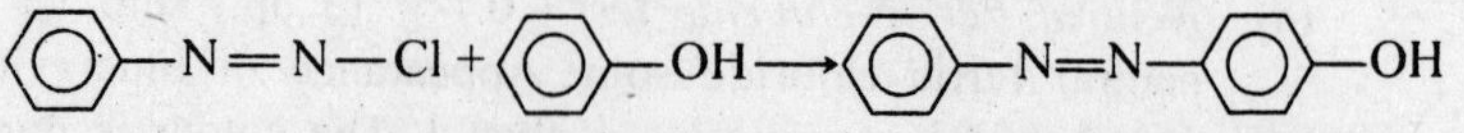

P - hydroxyazobenzene
(Orange dye)

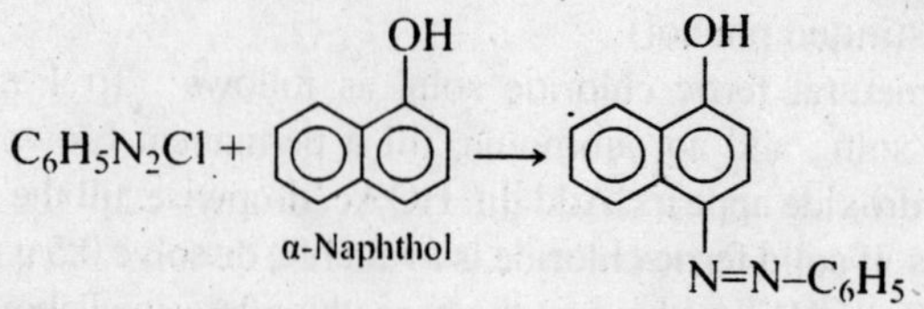

α-Naphthol

Scarlet-red dye

(e) Libermann's test : Mix 0.1 g sub. with 0.5 g $NaNO_2$ in a dry test tube. Heat for 2 mts. Cool. Add 5 dp. con. H_2SO_4. A blue or green soln. is got. Pour this into 30 ml water in a beaker, and add NaOH in excess. A red soln. obtained first : with excess NaOH, the colour changes to deep blue.

Note: All phenols do not give this test.)

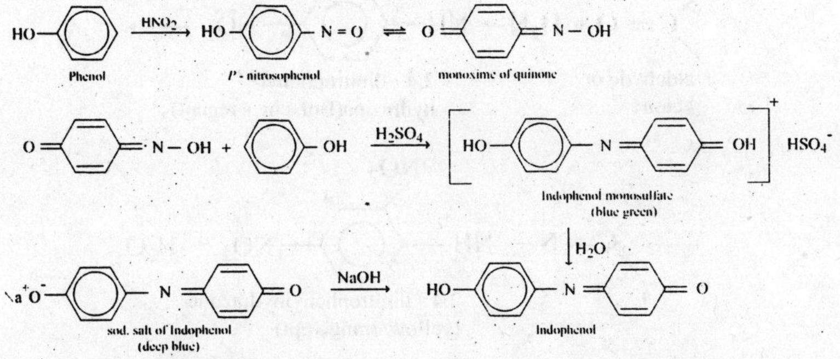

3. Tests for Esters

(*a*) Hydrolysis : 0.2 g. sub. + 2 ml 10 % NaOH. Boil for 3 mts. The disappearance of pleasant odour indicates ester.

$$\underset{\text{Methyl acetate}}{CH_3COOCH_3} + NaOH \longrightarrow \underset{\text{sod. acetate}}{CH_3COONa} + \underset{\text{methanol}}{CH_3OH}$$

(*b*) *Precipitation of acid* : Cool the reaction mixture from above test (a), add 1 ml HCl. A white ppt. indicates aromatic ester.

$$C_6H_5COONa + HCl \longrightarrow \underset{\substack{\text{Benzoic acid} \\ \text{(insoluble in } H_2O \text{ in cold)}}}{C_6H_5COOH\downarrow} + NaCl$$

(*c*) *Test with phenolphthalein* : 0.2 g. sub + 2 dp. phenolphthalein + 3 dp. NaOH soln. Heat. Pink colour disappears on heating; indicates an ester.

Initially the free NaOH gives pink colour with indicator On heating, NaOH is used for alkaline hydrolysis of ester: hence, free indicator (colourless) is liberated.

4. Tests for Aldehydes and Ketones

(*a*) *Borsche's test*: To 0.1 g sub, add 1 ml Borsche's reagent. Boil and add con. HCl. Cool. Add 1 ml water. Yellow or orange ppt. shows aldehyde or ketone.

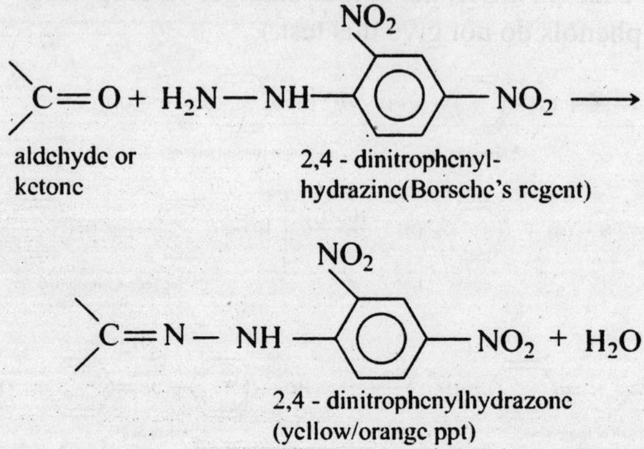

(*b*) *Phenylhydrazine test* : 0.1 g sub. + 1 ml ethanol + 1 ml phenylhydrazine. Heat in a boiling water bath for 10 mts. Yellow ppt. indicates aldehyde or ketone. Deep yellow osazone indicates aldoses (like glucose).

$$>C{=}O + \underset{\text{Phenylhydrazine}}{H_2N{-}NH{-}C_6H_5} \longrightarrow \underset{\substack{\text{Phenylhydrazone}\\ \text{(yellow ppt.)}}}{>C{=}N{-}NH{-}C_6H_5\downarrow} + H_2O$$

(*c*) *Semicarbazide test*: 0.2g sub. + a mixture of 0.2 g semicarbazide hydrochloride, 0.5 g sod. acetate and 2 ml water. Add about 1 ml of a alcohol, drop by drop, to get clear soln. Heat for 2 mts., cool. Colourless crystals of semicarbazone indicate aldehyde or ketone.

$$>C = O + \underset{\text{Semicarbazide}}{H_2N - NH - CO - NH_2} \longrightarrow \underset{\text{Semicarbazone}}{>C = N - NH - CO - NH_2} + H_2O$$

(d) *Schiff's reagent test* : (aldehydes only.) To 0.1 g sub, add 3–4 dp. Schiff's reagent. Shake well. Immediate pink or red colour shows aldehyde. (Some aliphatic ketones, like acetone, also give a pink colour, but very slowly).

Schiff's reagent is prepared from *p*-rosaniline hydrochloride.

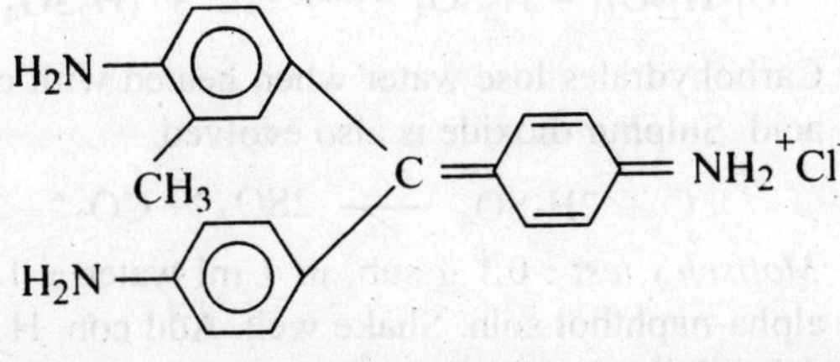

The pink colour of this salt soln. in water is decolourised by passing sulphur dioxide. This decolouration results from destruction of quinonoid structure of the dye by H_2SO_3. Aldehyde combines with H_2SO_3, restores the quinonoid structure, hence pink colour.

(*e*) *Fehling's soln. test* : (for aliphatic aldehydes only.) To 0.1 g. sub. add 1 ml each of Fehling's soln. "A" and "B". Heat to boil. A red ppt. shows aliphatic aldehyde or aldose (like glucose).
Fehling's soln. "A" : Copper sulfate soln.
Fehling's soln. "B" : Alkaline sod. pot. tartrate.

$$RCHO + 2CuO \longrightarrow RCOOH + \underset{\substack{\text{cuprous oxide} \\ \text{(red ppt.)}}}{Cu_2O\downarrow}$$

(*f*) *Tollen's reagent test* : (aldehydes only). To 0.1 g sub. add 1 ml Tollen's reagent. Heat the test tube in a water bath for 5–10 mts. Black ppt. (or bright silver mirror) indicates aldehyde or aldose (like glucose)
Tollen's reagent : Ammoniacal $AgNO_3$ soln.

$$RCHO + 2[Ag(NH_3)_2]OH \longrightarrow RCOONH_4 + \underset{\text{black ppt.}}{2Ag\downarrow} + 3NH_3 + H_2O$$

(*g*) *Legal's test* : (for ketones only) 0.1 g. sub + 2 ml water + 5dp. sod. nitroprusside soln. + 5 dp. NaOH + 5 dp. glacial HOAc. Orange colour changes to purple, indicates ketone.

5. Tests for Carbohydrates

(*a*) *Charring test* : To 0.1 g sub. in a dry test tube, add 5 dp. con. H_2SO_4. Heat gently. Charring (with burnt sugar smell) indicates carbohydrate.

$$C_{12}H_{22}O_{11} + H_2SO_4 \longrightarrow 12C + (H_2SO_4 + 11H_2O)$$

Carbohydrates lose water when heated with con. sulphuric acid. Sulphur dioxide is also evolved.

$$C + 2H_2SO_4 \longrightarrow 2SO_2\uparrow + CO_2\uparrow + 2H_2O$$

(*b*) *Molisch's test* : 0.1 g sub. in 1 ml water + 1 ml alcoholic alpha-naphthol soln. Shake well. Add con. H_2SO_4, drop by drop, through side of test tube. The formation of deep violet ring at the junction of the two liquids, and the spreading of colour on standing, shows carbohydrate.
The violet ring may be due to an unstable condensation product of alpha-naphthol with furfural (formed by the cyclodehydration of a carbohydrate with acid).

(*c*) *Osazone test* : 0.1 g. sub. in 3 ml water + 1 ml phenylhydrazine. Heat for 5–10 mts. in a boiling water bath. Yellow crystals of osazone show monosaccharide (like glucose).

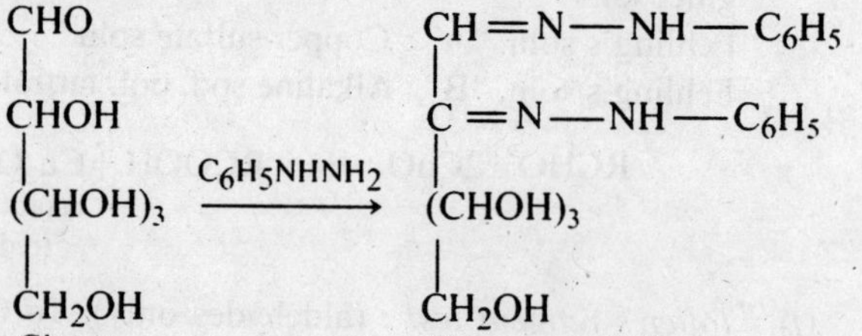

Glucose — Glucosazone (yellow)

(d) Refer test Nos. 4 (*e*) and (*f*) also, for monosaccharides (like glucose)

6. Tests for Alcohols

(a) *Na test* : 0.5 ml org. liquid + 1 ml benzene + small piece of sodium. Vigorous reaction with evolution of hydrogen (pop sound with burning splinter) indicates alcohol.

$$\underset{\text{alcohol}}{2ROH} + 2Na \longrightarrow \underset{\text{sod. alkoxide}}{2RONa} + H_2\uparrow$$

(b) *Esterification* : 0.1 g. sub. + 1 ml acetic anhydride +3 dp. con. H_2SO_4. Heat for 2-3 mts., cool, pour into 20 ml dil Na_2CO_3 soln.

$$ROH + (CH_3CO)_2O \longrightarrow \underset{\text{alkyl acetate}}{CH_3COOR} + CH_3COOH$$

HOAc is neutralised by Na_2CO_3 soln

TEST FOR N COMPOUNDS

***Note*:** Do the following tests, only if N had been detected in Na fusion test.)

7. Tests for Amines

(*a*) *Solubility test* : 0.1 g. sub. + 1 ml dil HCl. Shake well. If the sub. is soluble and gets reprecipitated on adding NaOH, an amine is indicated.

$$\underset{\text{aniline}}{C_6H_5NH_2} + HCl \longrightarrow \underset{\substack{\text{aniline}\\\text{hydrochloride}}}{C_6H_5\overset{+}{N}H_3\overset{-}{Cl}} \xrightarrow{NaOH} \underset{\substack{\text{aniline}\\\text{(regenerated)}}}{C_6H_5NH_2} + NaCl + H_2O$$

(*b*) *Carbylamine test* : 0.1 g. sub. + 3 dp. chloroform + 2 ml alc. KOH. Shake well. Warm gently. Offensive smell of isocyanide indicates primary amine (e.g., aniline)

$$C_6H_5NH_2 + CHCl_3 + 3KOH \rightarrow C_6H_5NC + 3KCl + 3H_2O$$

(*c*) *Dye test* : Dissolve 0.1 g. org. compd. in 3 ml dil. HCl. Cool (in tap or ice), add 1 ml sod. nitrite (0.5 g. / H_2O) soln. A scarlet red dye indicates a primary amine.

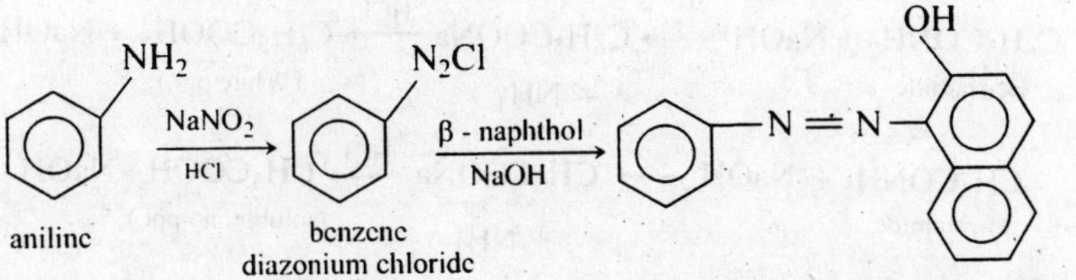

Scarlet - red dye

(*d*) *Acetylation* : 0.1 g. sub. + 1 ml glacial HOAc + 0.5 ml acetic anhydride. Heat gently for 3–5 mts. pour into 20 ml water in a beaker. White crystals indicate primary amine.

$$\underset{\text{aniline}}{C_6H_5NH_2} + CH_3COOH \xrightarrow{(CH_3CO)_2O} \underset{\text{acetanilide}}{C_6H_5NHCOCH_3} + H_2O$$

(*e*) *Libermann's test* : Dissolve 0.1 g. sub. in 3 ml ethyl alcohol, add 1 ml con. HCl. Cool (in tap or ice), add 1 ml $NaNO_2$ soln. Add 3 ml ether, shake well and separate ether layer and evaporate it carefully. Add 5 dp. phenol to the residue. Warm, cool, and 5 dp. con. sulphuric acid. Pour the product into 20 ml of dil NaOH soln. First a red and then a blue soln. indicates a secondary amine.

$$\underset{\text{N - methyl aniline}}{C_6H_5NHCH_3} + \underset{\text{Nitrous acid}}{HONO} \longrightarrow \underset{\text{N - Nitrosamine}}{C_6H_5-N(-N{=}O)-CH_3}$$

(*f*) *Quaternary salt test* : 0.1 g. sub. + 1 ml CH_3I soln. Shake well. Heat gently, allow to stand for 2–5 mts. white crystaline solid (quaternary salt) indicates tertiary amine.

$$(CH_3)_3N + CH_3I \longrightarrow \underset{\text{Quaternary salt}}{(CH_3)_4N^+I^-}$$

8. Tests for Amides

(*a*) *NaOH test*: 0.1 g. sub. + 2 ml 10% NaOH soln. Heat strongly. Smell of NH_3 indicates amide. Heat till no more NH_3 is evolved. Add 5 dp. con. HCl. White ppt. indicates aromatic amide; no ppt. indicates aliphatic amide.

$$\underset{\text{Benzamide}}{C_6H_5CONH_2} + NaOH \longrightarrow C_6H_5COONa + NH_3 \xrightarrow{HCl} \underset{\text{(White ppt.)}}{C_6H_5COOH\downarrow} + NaOH$$

$$\underset{\text{acetamide}}{CH_3CONH_2} + NaOH \longrightarrow CH_3COONa + NH_3 \xrightarrow{HCl} \underset{\text{(soluble, no ppt.)}}{CH_3COOH + NaOH}$$

(*b*) *Urea nitrate test*: 0.1 g. sub. + 1 ml H_2O + 1 ml con. HNO_3. Shake well. White ppt. shows a diamide (like urea).

$$\underset{\text{urea}}{NH_2CONH_2} + HNO_3 \longrightarrow \underset{\text{urea nitrate}}{CO(NH_2)_2 \,.\, HNO_3\downarrow}$$

(*c*) *Urea oxalate test*: Shake 0.1 g. sub. with 3 ml satd $H_2C_2O_4$ soln. White ppt. shows a diamide.

$$\underset{\text{urea}}{2CO(NH_2)_2} + H_2C_2O_4 \longrightarrow \underset{\text{urea oxalate}}{[CO(NH_2)_2]_2 \,.\, H_2C_2O_4\downarrow}$$

(*d*) *Biuret test* : Heat 0.2 g. sub. strongly in a dry test tube till no more NH_3 is evolved. Cool. Add 1 ml H_2O and 5 dp. $CuSO_4$ soln.and 3 dp. NaOH soln. Violet colour shows diamide.

$$NH_2CONH_2 + NH_2CONH_2 \longrightarrow \underset{\text{Biuret}}{NH_2-CO-NH-CO-NH_2} + NH_3$$

$$\downarrow NaOH, CuSO_4$$

violet colour

9. Tests for Nitro group

(a) *Mulliken and Barker's test* : Dissolve 0.1 g. sub. in 2ml ethanol. Add 0.5 g. NH_4Cl and 0.5 g. zinc dust. Heat for 2 mts. Allow to stand for 3 mts. Filter. Divide the filtrate into two parts.

$$C_6H_5NO_2 \xrightarrow{Zn/NH_4Cl} \underset{\text{Nitrosobenzene}}{C_6H_5NO} \longrightarrow \underset{\text{Phenyl hydroxylamine}}{C_6H_5NHOH}$$

(i) To one part, add 2 ml Tollen's reagent. Heat in water bath. Black ppt. or bright Ag mirror indicates nitro group.

$$C_6H_5NHOH + 2[Ag(NH_3)_2]OH \longrightarrow C_6H_5NO + 2Ag\downarrow + 4NH_3\uparrow + 2H_2O$$

(ii) To other part, add 1 ml each of Fehling's soln. "A" and "B". Heat for 2mts. A red ppt. (cuprous oxide) indicates nitro compd.

$$C_6H_5NHOH + 2CuO \longrightarrow C_6H_5NO + \underset{\text{red}}{Cu_2O\downarrow} + H_2O$$

(b) Reduction : To 0.3 g. sub. add. 5 ml con. HCl and a piece of Sn. Heat for 3–5 mts. Filter. Add KOH soln. till it becomes alkaline. An amine separates as a liquid (which may be tested by reactions under amines) : shows presence of nitro compd.

$$C_6H_5NO_2 \xrightarrow{Sn/HCl} C_6H_5NH_2$$

10. Tests for Anilides

(a) *Carbylamine test* : 0.1 g. sub. + 2 ml 10 % NaOH soln. Heat for 2–3 mts. Add 2 dp. chloroform. Heat. Unpleasant smell (isocyanide) shows anilide.

$$\underset{\text{acetanilide}}{C_6H_5NHCOCH_3} \xrightarrow{NaOH} \underset{\text{aniline}}{C_6H_5NH_2} + CH_3COONa$$

$$C_6H_5NH_2 + CHCl_3 + 3NaOH \longrightarrow \underset{\text{Phenyl isocyanide}}{C_6H_5NC} + 3NaCl + 3H_2O$$

(b) *Bromination* : Dissolve 0.1 g. sub. in 3 ml glacial HOAc. Add 1 ml bromine in HOAc. Shake well, pour into 10 ml water. Pale yellow ppt. indicates anilide.

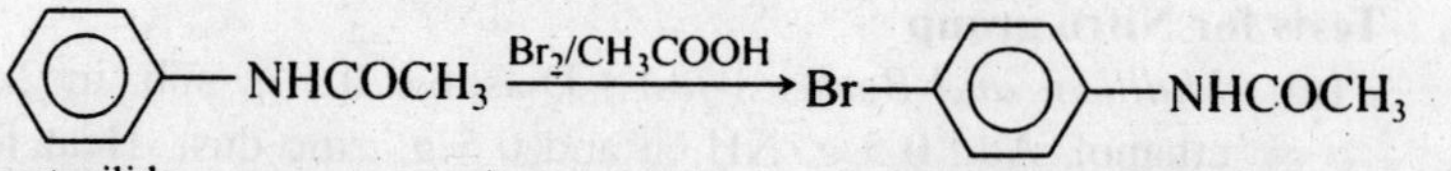

acetanilide p-bromoacetanilide

(b) Dye test : Heat 0.1 g. sub. with 3 ml con. HCl for 2–3 mts. Cool the soln. (in tap or ice). Add slowly soln. of 0.5 g. sodium nitrite in water + soln. of 0.5 g. β-naphthol in 2 ml NaOH. Scarlet red dye indicates anilide.

$$C_6H_5NHCOCH_3 + HCl \longrightarrow C_6H_5NH_2 + CH_3COCl$$

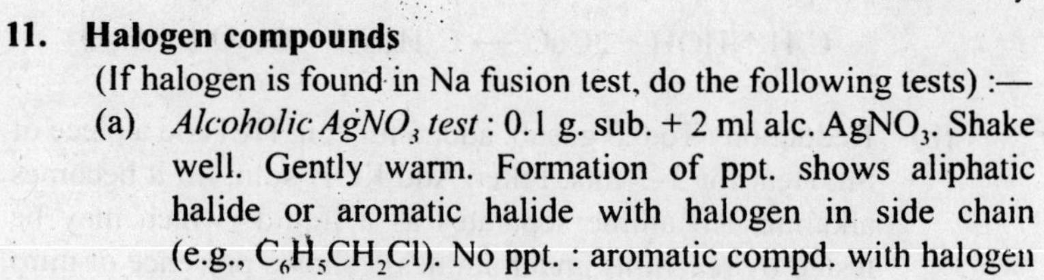

Scarlet - red dye

11. Halogen compounds

(If halogen is found in Na fusion test, do the following tests) :—

(a) *Alcoholic $AgNO_3$ test* : 0.1 g. sub. + 2 ml alc. $AgNO_3$. Shake well. Gently warm. Formation of ppt. shows aliphatic halide or aromatic halide with halogen in side chain (e.g., $C_6H_5CH_2Cl$). No ppt. : aromatic compd. with halogen in nucleus.

(b) *Anilide test* : 0.1 g. sub. + 1 ml aniline. Shake vigorously, pour it into 10 ml water. White ppt. indicates acid chloride.

$$\underset{\text{Acetyl chloride}}{CH_3COCl} + C_6H_5NH_2 \longrightarrow \underset{\text{Acetanilide}}{C_6H_5NHCOCH_3\downarrow} + HCl$$

12. Test for Sulphonic acid group

(If S is found in Na fusion test, do the following test) :—

Barium chloride test : Burn 0.1 g. sub. in a Ni spatula. Dissolve residue in dil. HCl, add barium chloride soln. White ppt. indicates sulphonic acid (e.g., $C_6H_5SO_2OH$ – benzene sulphonic acid):

It is possible that the sulphonic acid forms, on decomposition, H_2SO_4, which gives a white ppt. of $BaSO_4$ with $BaCl_2$.

13. Test for Hydrocarbons

If the given unknown organic compound does not answer for any of the above functional groups, it can be considered to be a hydrocarbon.

(a) *Test for Aromatic hydrocarbon*: Dissolve 30 mg. (or 3 drops) of the substance in 1 ml of chloroform. In a small dry test tube, heat a small lump of fresh anhydrous $AlCl_3$ over a tiny flame, until it slowly sublimes and forms a thin coating over the sides of the tube. Now add the chloroform solution of the substance, and roll the test tube till the sublimate gets wetted completely.

An intense colour (yellow or greenish blue) develops, and this colour remains unchanged for a long time. This indicates the presence of an aromatic hydrocarbon (Friedel – Crafts reaction takes place).

***Note*:** Benzene gives a deep yellow colour ; naphthalene, – blue; diphenyl, – bluish – green.

(b) *Preparation of Picrate Derivative* : A picrate derivative of an aromatic hydrocarbon can be prepared as follows : To about 0.5 g. (or 10 drops) of the given substance, 0.5 g. of picric acid and 2 ml of rectified spirit are added, and shaken well. The picrate of the given hydrocarbon separates out as a yellow solid. It is filtered, dried and the m.p. is determined.

***Note*:** Picrate of benzene cannot be prepared; picrates of solid hydrocarbons like naphthalene, anthracene and phenanthrene can readily be prepared. For benzene and substituted benzenes, solid dinitro-derivatives can be prepared : *see* nitration procedure in F : 1 (c).

F. Derivatives

After identifying functional group/s, confirm by preparing a solid derivative as described below :—

1. Carboxylic acid

(a) *Anilide derivative* : Mix 1 g. sub. with 5 ml aniline in boiling test tube. Add a porcelain bit, fit with air condenser and heat for 30 mts. in a sand-bath. Pour hot mixture into 50 ml dil. HCl, stir well. Allow mixture to stand for 10–15 minutes. White ppt. (anilide) separates out. Filter, wash with cold water, recrystallise from alcohol.

$$\underset{\text{benzoic acid}}{C_6H_5COOH} + \underset{\text{aniline}}{C_6H_5NH_2} \longrightarrow \underset{\text{benzanilide}}{C_6H_5NHCOC_6H_5} \downarrow + H_2O$$

(b) *S-Benzyl isothiuronium derivative* : Dissolve 1 g. sub. in 10 ml 10 % NaOH, and 10 ml of 1 g. S-benzyl isothiuronium chloride in NaOH. Shake well. White ppt. obtains. Filter wash with cold water.

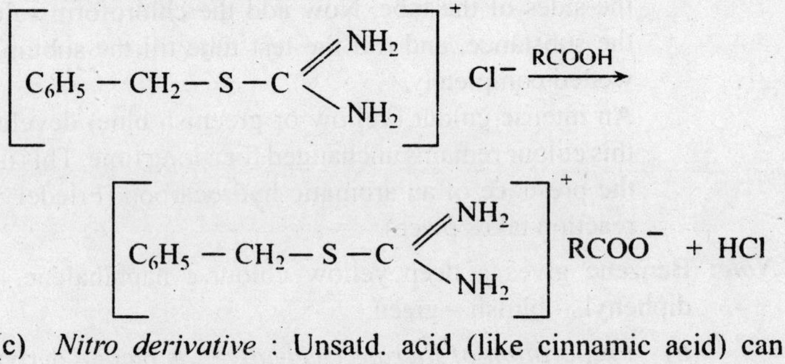

(c) *Nitro derivative* : Unsatd. acid (like cinnamic acid) can be nitrated using fuming nitric acid. Mix 1 g. sub. + 2 ml fuming nitric acid slowly with constant stirring in dry test tube. Keep aside for 10–15 mts. Pour into 50 ml distilled water in beaker. Yellow crystals separate out. Filter at pump, wash with water, recrystallise from alcohol.

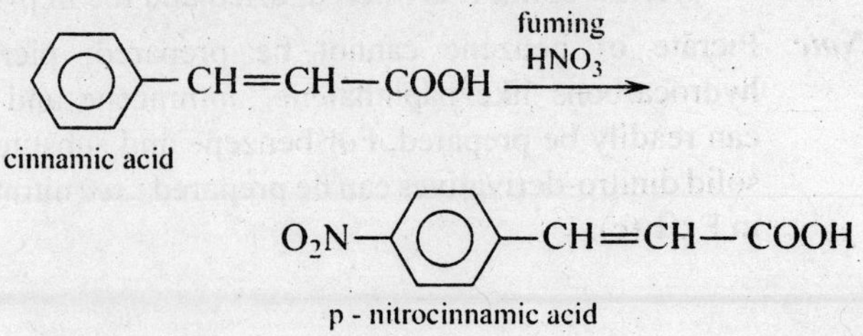

(d) *Bromo derivative* : For aromatic acid (like salicyclic acid and cinnamic acid) bromo derivative can be prepared. Mix about 1 g. sub. with 5 ml Br_2 in glacial HOAc slowly, with constant stirring, till yellow colour develops. Shake well for 10–15 mts pour it into 50 ml water with stirring. Filter yellow crystals wash with water, recrystallise from dil. alcohol.

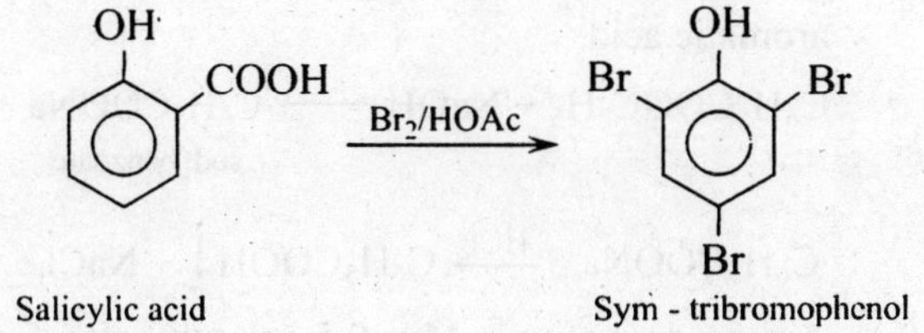

2. Phenol :

(a) *Bromo-derivative* : [Refer (d) above]

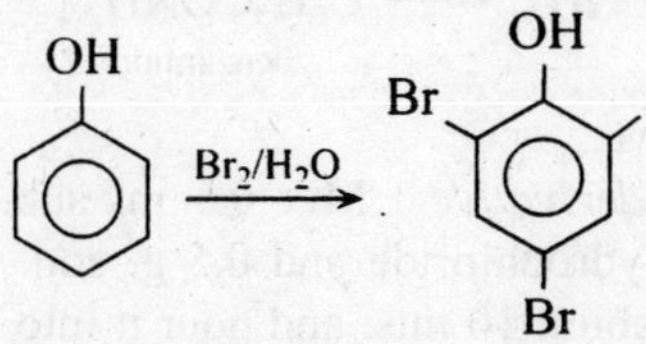

2,4,6 - tribromophenol

(b) *Nitro derivative* : Mix 1 g. sub. with 2 ml con. H_2SO_4 warm slightly (2 mts.). Cool, add 2 ml con. HNO_3 drop by drop, Heat (10 mts.), pour into 20 ml water (beaker). Yellow solid shows phenolic group. Recrystallise small portion from boiling water.

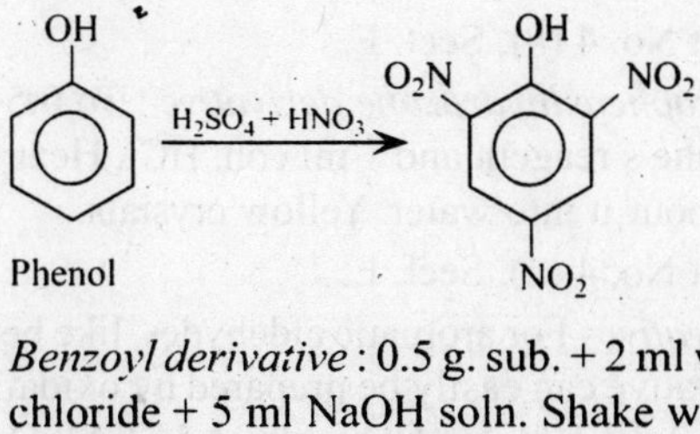

Picric acid

(c) *Benzoyl derivative* : 0.5 g. sub. + 2 ml water + 0.5 ml benzoyl chloride + 5 ml NaOH soln. Shake well. (5 mts.), pour into

20 ml water (beaker). White ppt. (benzoyl derivative of phenol).

$$C_6H_5OH + C_6H_5COCl \xrightarrow{NaOH} C_6H_5OCOC_6H_5\downarrow + HCl$$

3. Esters

(a) *Acid derivative* : To 0.5 ml sub., add 10 ml 10 % NaOH soln. Heat for about 2–4 mts. Add con. HCl. White ppt. shows an aromatic acid.

$$C_6H_5COOC_2H_5 + NaOH \longrightarrow \underset{\text{sod. benzoate}}{C_6H_5COONa} + C_2H_5OH$$

$$C_6H_5COONa \xrightarrow{HCl} C_6H_5COOH\downarrow + NaCl$$

(b) Amide derivative : Mix 0.5 ml sub. with 5 ml liq. NH_3. Shake well, and leave for 10 mts. The formation of a white ppt. shows an amide.

$$C_6H_5COOCH_3 + NH_3 \longrightarrow \underset{\text{benzamide}}{C_6H_5CONH_2\downarrow} + CH_3OH$$

4. Aldehydes and Ketones

(a) *Semicarbazone derivative* : Mix 0.5 ml sub. with 0.5 g. semicarbazide hydrochloride and 0.5 g. sod. acetate soln. Heat gently for about 10 mts. and pour it into water. White crystals of semicarbazone is obtained.
Refer Test No. 4 (*c*), Sect. E.

(b) *Phenylhydrazone derivative* : To 0.5 ml sub., mix 1 ml phenylhydrazine and 1 ml glacial HOAc. Heat gently for about 10 mts. Pour it into water. Yellow crystals show the derivative of carbonyl compounds.
Refer Test No. 4 (*b*), Sect. E.

(c) 2,4–*dinitrophenylhydrazone derivative* : To 0.5 g. sub., add 1 ml Borche's reagent and 1 ml con. HCl. Heat for about 10 mts. and pour it into water. Yellow crystals.
Refer Test No. 4 (*a*), Sect. E.

(d) *Acid derivative* : For aromatic aldehydes, like benzaldehyde, acid derivative can easily be prepared by oxidation. Mix 1g. sub. with 2 ml 5 % Na_2CO_3 soln. Add $KMnO_4$ soln., drop by drop, till the colour remains. Heat for about 10 mts. Add dil. sulphuric acid to the hot soln. Then, add a strong soln. of sod. sulphite till the soln. becomes colourless. On cooling, white crystals separates out.

$$\underset{\text{benzaldehyde}}{C_6H_5CHO} \xrightarrow{\text{alk. } KMnO_4} \underset{\text{benzoic acid}}{C_6H_5\text{–}COOH} \downarrow$$

5. Carbohydrates

Phenylhydrazone derivative : Refer 4 (b) derivative.

6. Alcohols

Acid derivative : Refer 4 (d) derivative.

7. Amines

(a) *Benzoyl derivatives* : Refer 2 (c) derivative.

$$C_6H_5NH_2 + C_6H_5COCl + NaOH \longrightarrow C_6H_5NHCOC_6H_5 + NaCl + H_2O$$

(b) *Acetyl derivative* : Mix 0.5 g. sub.with 1 ml acetic anhydride and 1 ml. glacial HOAc. Heat for about 10 mts. pour it into 20 ml water. White crystals.

$$C_6H_5NH_2 + (CH_3CO)_2O \longrightarrow \underset{\text{acetanilide}}{C_6H_5NHCOCH_3} \downarrow + CH_3COOH$$

8. Amides

Acid derivative : Refer 3 (a) derivative.

$$\underset{\text{benzamide}}{C_6H_5CONH_2} + NaOH \longrightarrow \underset{\text{sod. benzoate}}{C_6H_5COONa} + NH_3$$

$$\downarrow HCl$$

$$\underset{\text{benzoic acid}}{C_6H_5COOH} + NaCl$$

9. Nitro group

Reduce the nitro compounds to the corresponding amines by heating with tin and HCl. Converts the amine into acetyl or benzoyl derivative. Refer 7 (*a*) and (*b*) derivative.

10. Anilides

(a) *Bromo derivative* : Refer 1 (*d*) derivative.

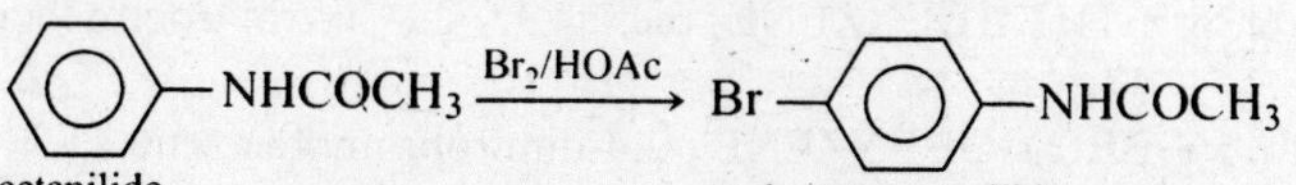

acetanilide p - bromoacetanilide

11. Halogen compounds

(*a*) *Acid derivative* : Refer 3 (a) derivative.

$$C_6H_5CH_2Cl \xrightarrow{NaOH} C_6H_5CH_2OH \xrightarrow{(O)} C_6H_5COOH$$

(*b*) *Nitro derivatives* : Aromatic halogen compounds (with halogen in the nucleus) can be nitrated to get nitro-derivatives Refer 2 (*b*) derivative.

G. Physical constants of some organic compounds and their derivatives

A list of common organic substances is given below, in the order of the increasing boiling / melting points (in ºC). For each substance, derivative that can be readily prepared are also given with their m.p.

Use 0–360º C. thermometer for m.p. / b.p. determination. For m.p., use a coconut oil-bath.

B.P. LIQUIDS

56 ACETONE : iodoform 119, 2,4-dinitrophenylhydrazone 128, semicarbazone 190, oxime 59.

77 *n*- BUTYL AMINE : picrate 151.

78 ETHYL ALCOHOL : iodoform 119, 3, 5-dinitrobenzoate 94.

80 METHYL ETHYL KETONE: iodoform 119, 2,4-dinitrophenylhydrazone 115, semicarbazone 146.

80 BENZENE : *m*-dinitrobenzene 90.

82 ISO–PROPYL ALCOHOL: iodoform 119; 3,5-dinitrobenzoate 122.

97 *n*-PROPYL ALCOHOL: 3,5-dinitrobenzoate 75.

102 CROTONALDEHYDE: 2,4-dinitrophenyylhydrazone 190, semicarbazone 199.

110 TOLUENE: benzoic acid 121, 2, 4-dinitrotoluene 70.

118 ACETIC ACID: anilide 114, amide 82.

132 CHLOROBENZENE : 2,4-dinitrochlorobenzene 52.

135 ETHYL BENZENE : picrate 97, 2, 4, 6-trinitroethylbenzene 37.

156 BROMOBENZENE : 2,4-dinitrobromobenzene 75.

176 *o*-CHLOROPHENOL : dinitrobenzoate 143.

179 BENZALDEHYDE : benzoic acid 121, anilide 117.
179 BENZALDEHYDE : benzoic acid 121, phenylhydrazone 157, 2,4-dinitrophenylhydrazone, 235, semicarbazone 224.
180 *ortho* - DICHLOROBENZENE : 4,5-dinitrochlorobenzene 110.
181 PHENOL : benzoate 69, 2,4,6-tribromophenol 93, aryloxyacetic acid 99, picric acid 122.
183 ANILINE : acetyl deriv. 114, benzoyl deriv. 162, 2,4,6-tribromoaniline 118.
185 BENZYL AMINE : picrate 196.
190 *o*-CRESOL : aryloxyacetic acid 152.
193 METHYLANILINE : acetyl deriv, 102, benzoyl deriv. 63.
196 SALICYLALDEHYDE : phenylhydrazone 143, semicarbazone 230, 2,4–dinitrophenylhydrazone 252.
198 METHYL BENZOATE : benzoic acid 121.
198 BENZYL BROMIDE : benzoic acid 121, anilide 177.
199 *o*-TOLUIDINE : acetyl deriv. 112, benzoyl deriv. 143. picrate 212.
201 *p*-CRESOL : benzoate 71, aryloxyacetic acid 136.
202 *m*-CRESOL : benzoate 54, aryloxyacetic acid 103, tribromo deriv 84, trinitro deriv. 105.
202 ACETOPHENONE : semicarbazone 119, 2,4, dinitro-phenyl hydrazone 250, oxime 59.
203 *m*-TOLUIDINE : acetyl deriv. 66, benzoyl deriv. 125.
205 BENZYL ALCOHOL : benzoic acid 121.
207 *o*-CHLOROANILINE : acetyl deriv. 87, benzoyl deriv. 99.
210 NITROBENZENE : m-dinitrobenzene 90.
213 ETHYL BENZOATE : benzoic acid 121.
216 *o*-NITROPHENOL : acetyl deriv. 41, benzoate 59.
224 METHYL SALICYLATE : salicylic acid 156.
231 ETHYL SALICYLATE : salicylic acid 156.
250 *o*-BROMOANILINE : acetyl deriv. 99, benzoyl deriv. 116.
261 METHYL CINNAMATE : cinnamic acid 133.
273 ETHYL CINNAMATE : cinnamic acid 133.
298 DIETHYL PHTHALATE : phthalic acid 195.

M.P. SOLIDS

44 *o*-NITROPHENOL : benzoate 59, aryloxyacetic acid 158.

45 *p*-TOLUIDINE : acetyl deriv. 148, benzoyl deriv. 158.

48 BENZOPHENONE : semicarbazone 164, oxime 141.

50 α-NAPTHYLAMINE : acetyl deriv. 159, benzoyl deriv. 161.

66 *p*-BROMOANILINE : acetyl deriv. 167, benzoyl deriv. 204.

68 PHENYLBENZOATE : benzoic acid 121.

70 DIPHENYL : 4,4-dinitrodiphenyl 234, dibromodiphenyl 164.

80 NAPHTHALENE : Picrate 149, 1-nitronaphthalene 61.

81 VANILLIN : semicarbazone 239, oxime 117, phenylhydrazone 105.

82 ACETAMIDE.

94 α-NAPHTHOL : picrate 149, benzoate 56.

95 BENZIL : oxime 237, semicarbazone 244, phenylhydrazone 235.

95 *m*-NITROPHENOL : benzoate 95.

97 OXALIC ACID : anilide 246.

101 FRUCTOSE : osazone 205.

102 RESORCINOL : benzoate 117.

112 β - NAPHTHYLAMINE : acetyl deriv. 134, benzoyl deriv. 162, picrate 195.

114 *m*-NITROANILINE: acetyl deriv. 155.

114 ACETANILIDE: *p*-bromoacetanilide 167, p-nitro deriv. 214.

114 *p*-NITROPHENOL: benzoate 142.

121 BENZOIC ACID: anilide 262, *p*-toluidide 158.

122 β-NAPHTHOL: benzoate 107, methyl ether 72, picrate 156.

126 SUCCINIMIDE : succinic acid 185.

128 BENZAMIDE : benzoic acid 121.

132 UREA : nitrate 163, oxalate 170.

133 PYROGALLOL : triacetate 170.

133 CINNAMIC ACID : anilide 153, *p*-toluidide 168, *p*-nitro deriv. 287.

137 BENZOIN : semicarbazone 206, phenylhydrazone 235.

141 *m*-NITROBENZOIC ACID : anilide 154, amide 142.

146 GLUCOSE : osazone 205.

146 ANTHRANILIC ACID (*o*-aminobenzoic acid) : acetyl deriv. 185.
148 *p*-NITROANILINE : acetyl deriv. 215, benzoyl deriv. 198.
158 SALICYLIC ACID : anilide 229, *p*-toluidide 234, acetyl deriv. 136, 5-nitrosalicylic acid 220.
183 2,4-DINITROBENZOIC ACID : amide 204.
185 SUCCINIC ACID : anilide 229, *p*-toluidide 234.
195 PHTHALIC ACID : anilide 204.
216 ANTHRACENE : picrate 138, anthraquinone 286.
232 GLYCINE : benzoate 187, 3,5-dinitrobenzoate 179.
233 PHTHALIMIDE : phthalic acid 195.
280 SULPHANILIC ACID (decomposes) : amide 164.

H. PREPARATION OF SPECIAL REAGENTS

1. *Alcoholic potash* : 30 % KOH in rectified spirit. Allow soln. to settle (for 24 hours). Decant clear soln.
2. *Alcoholic silver nitrate solu*tion : 2% soln. of $AgNO_3$ in rectified spirit. Decant clear soln.
3. *Borsche's reagent* : 1% soln. of 2,4 - dinitrophenyl hydrazine in rectified spirit (or methyl alcohol). Heat till clear soln. appears. Cool, filter.
4. *Bromine in* CCl_4 : 4 ml liq. bromine in 100 ml CCl_4.
5. *Bromine water* : Saturated soln. of Br_2 in water.
6. *Bromine in HOAc* : 4% soln. (1 ml Br_2 in 75 ml HOAc).
7. *Fehling's solution*–A : 7% $CuSO_4$ in water.
8. *Fehling's solution* –B : 35 g. Rochelle salt and 15 g. NaOH dissolved separately in 40 ml of water. The solns. mixed. Volume made up to 800 ml (1 ml of soln. A and 1 ml of soln. B mixed for test.)
9. *Neutral* $FeCl_3$ *solution* : To 1 ml ferric chloride soln. add aq. ammonia till permanent brown ppt. appears. Add dil. HOAc, drop by drop, till ppt, just dissolves.
10. *Phenylhydrazine reagent* : 10 ml phenylhydrazine + 10 ml glacial HOAc + 30. ml water.
11. *Potassium permanganate solution* : 2% $KMnO_4$ in water.
12. *Schiff's reagent* : 1 g *p*-rosaniline hydrochloride dissolved in 100 ml water, decolourised by passing sulphur dioxide. Soln. is filtered, diluted to 500 ml with distilled water, kept

in brown bottle.

13. *Sod. bisulfite soln.* : Satd. soln. of $NaHSO_3$ in water.
14. *Sod. nitroprusside soln* : 1% soln. in water.
15. Tollen's reagent—A : 10 g. silver nitrate dissolved in 100 ml of dilute aq. ammonia.
16. Tollen's reagent—B : 10 g. of NaOH dissolved in 100 ml water. (When required, 1 ml. A mixed with 1 ml B, used for test)

I. ANALYSIS OF MIXTURES

The identification of the components of a mixture of organic substances without separating them is a difficult task. This is due to the presence of two or more functional groups, which may function independent of each other. The separation of the components from a given mixture mainly depends on their solubility in suitable solvents, and their chemical properties.

Classification of substances in a mixture : The components present in given mixture can be classified into the following four categories :—

(*a*) Acidic substances;
(*b*) Phenolic substances;
(*c*) Basic substances; and
(*d*) Neutral substances.

The different combinations of organic substances than can be given in a mixture are listed below :—

(*a*) Acid sub. + neutral sub.
(*b*) Basic sub. + neutral sub.
(c) Phenolic sub. + neutral sub.
(*d*) Acid sub. + phenolic sub.
(*e*) Phenolic sub. + basic sub.

Acidic substances include carboxylic acids, sulphonic acids and rarely imides.

Phenolic substances are weakly acidic; they include phenol, substituted phenols and naphthols.

Basic substances include amino compounds and cyclic bases.

Neutral substances are generally classified into more reactive and less reactive substances. Alcohols, aldehydes, ketones, esters and carbohydrates contain more reactive functional groups. Among the less reactive compounds are the hydrocarbons and aromatic nitro compounds.

Pilot Tests

Before effecting the separation of the components from the bulk of the given mixture, the method of separation must be determined. This is usually done by subjecting a small amount of the mixture to the following tests (called pilot tests) :–

(*i*) *Solubility in water* : About 0.2 g. of the mixture is treated with 2 ml of water. The mixture is shaken well, noting whether a part or the whole of the mixture has dissolved. The clear solution is then tested with litmus, to find out whether the dissolved component is acidic or basic. If a component is soluble in water, its regeneration in a solid form is not possible.

(*ii*) *Solubility in ether* : About 0.2 g. of the mixture is treated with 2 ml of ether. The mixture is shaken well, noting whether a part of the whole of the mixture has dissolved.

Three cases may be observed :

(*a*) Both components of the mixture are insoluble in ether.

(*b*) Only one component of the mixture is soluble in ether.

(c) Both components of the mixture are soluble in ether.

Case (*a*) : *Both components insoluble in ether*

A mixture of this type is very rare. Only a very few mixtures can be made, where both components are insoluble in ether. Such a mixture combination is not recommended. The separation of the components from this mixture is beyond the scope of this lab. manual.

Case (*b*) : *Only one component soluble in ether*

Separation of the components from a mixture of this type is easy. The insoluble components is filtered off, and the soluble component can be recovered by the evaporation of the ether layer. This method of separation is called *ether separation*.

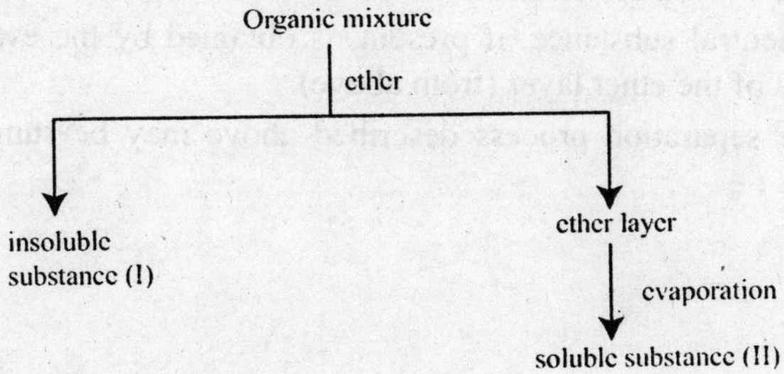

Case (*c*) : *Both components soluble in ether*

Organic mixtures of this type are common. Here, the components can be separated on the basis of their chemical behaviour. The mixture is treated with the following reagents in order : aq. sodium bicarbonate, aq. NaOH and dil. HCl. The tests are carried out as follows :—

The organic mixture which has dissolved in ether, is treated with 5 *ml.* of *aq. $NaHCO_3$* in a test-tube. The mixture is shaken well, and then allowed to stand for 1 minute. The two layers obtained are separated using a dropper. The bottom aq. layer is then treated with 2 ml of con HCl. The appearance of a precipitate indicates the presence of an aromatic acid. If no regeneration takes place, it indicates the absence of an acidic component in the mixture.

Note: Aliphatic acids (like oxalic acid and succinic acid) cannot be regenerated as solids, due to their solubility in water. Hence, if no precipitate appears after the neutralisation with con. HCl, then about 10 ml of ether is added to the aqueous solution and shaken well. The top ether layer is separated, and evaporated to dryness. An aliphatic acid will be obtained as a solid.)

The ether layer obtained from the above is then treated with 5 ml, *of aq. NaOH*. The mixture is shaken well, and allowed to stand for one minute. The bottom aqueous layer is separated, using a dropper, and treated with 2 ml of con. HCl. A regeneration of dissolved substance indicates the presence of a phenol. (A phenolic compound is considered as absent, if there is no regeneration on acidification of aq. layer).

The ether layer obtained above is then treated with 3 ml *of dil. HCl.* the mixture is shaken well, and allowed to stand for one minute. The bottom aq. layer is separated using a dropper, and it is neutralised by adding 3 ml of aq.NaOH. A regeneration indicates the presence of a basic (amine) component. (The absence of regeneration indicates the absence of a basic component.)

A neutral substance, if present, is obtained by the evaporation to dryness of the ether layer (from above).

The separation process described above may be summarised, as follows :-

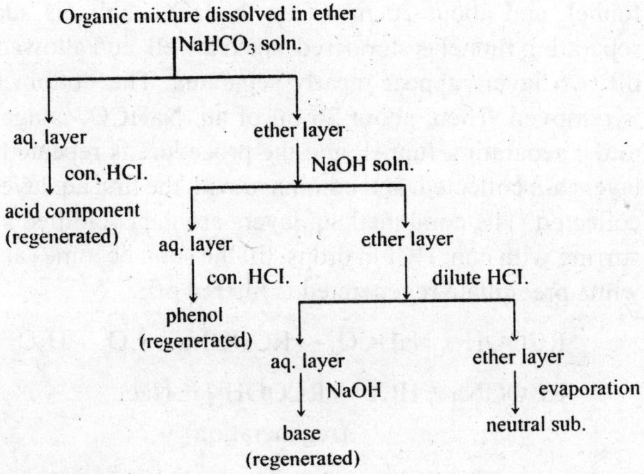

Bulk Separation

After determining the separation procedure by doing the pilot tests outlined above, the bulk (major portion of the mixture) is taken, and the components are separated as follows :–

1. *Ether separation* (only one component is soluble in ether): About 4-5 g. of the given mixture (bulk) is shaken well with 10 ml of ether. The clear ether layer is decanted. This procedure is repeated three times, to make sure that the ether-soluble component is completely dissolved (or extracted) from the mixture. This ether-soluble component is recovered by the evaporation to dryness of the combined ether layers.

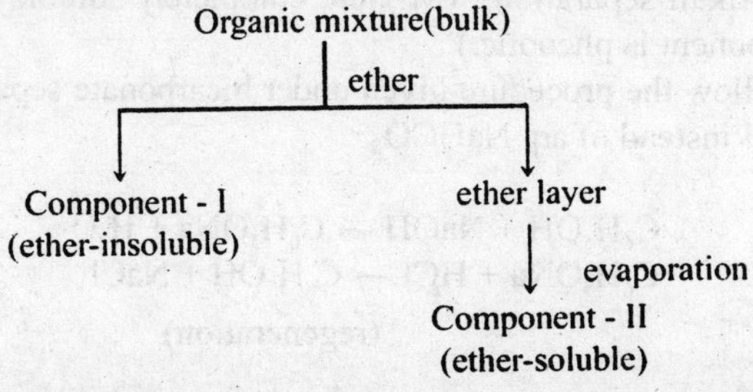

2. *Bicarbonate separation* : (Mixture completely soluble in ether, and one component is acidic.)
   About 4-5 g. of the organic mixture (bulk) is first dissolved in the minimum amount of ether. It is then taken in a 250 ml separating

funnel, and about 20 ml of aq. $NaHCO_3$ soln. is added. The separating funnel is stoppered, shaken well, and allowed to stand till two layers appear clearly separated. The bottom aq. layer is removed. Then, about 20 ml of aq. $NaHCO_3$ is again added to the separating funnel, and the procedure is repeated. The aq. layers are collected, and combined with the first aq. layer already collected. The combined aq. layers are then acidified slowly by stirring with con. HCl in drops, till the soln. becomes acidic. The white precipitate regenerated is filtered off.

$$RCOOH + NaHCO_3 \rightarrow RCOONa + CO_2 + H_2O$$

$$RCOONa + HCl \rightarrow \underset{\text{(regeneration)}}{RCOOH \downarrow} + Nacl$$

The ether layer is evaporated to dryness to recover the component-II

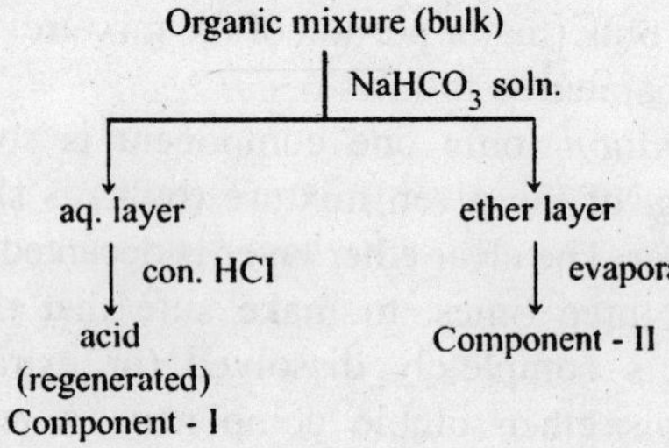

3. Alkali separation : (Mixture completely soluble in ether, and one component is phenolic.)

Follow the procedure given under bicarbonate separation, using aq. NaOH instead of aq. $NaHCO_3$

$$C_6H_5OH + NaOH \rightarrow C_6H_5ONa + H_2O$$

$$C_6H_5ONa + HCl \rightarrow C_6H_5OH + NaCl$$

(regeneration)

The ether layer from the separating funnel, on evaporation, gives mponent-II.

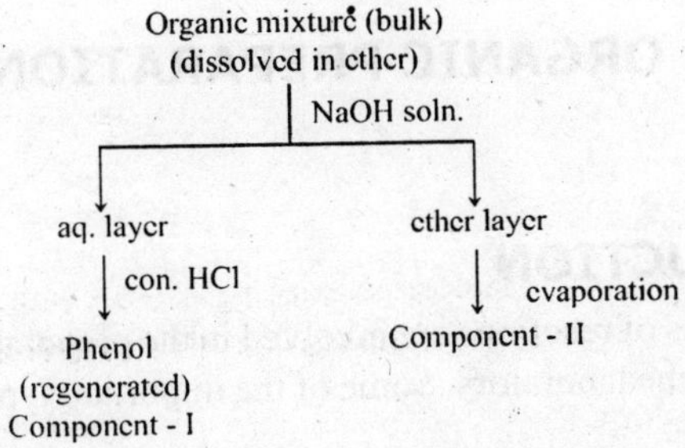

4. *Acid separation:* (Mixture competely soluble in ether, and one component is basic.)

About 4-5 g. of the organic mixture (bulk) is first dissolved in the minimum amount of ether. It is then taken in a 250 ml separating funnel, and about 20 ml of dil. HCl is added. The separating funnel is stoppered, shaken well and allowed to stand for 5 minutes. The bottom aq. layer is removed. Then about 20 ml of dil. HCl is added to the separating funnel and the procedure is repeated. All the aq. layer are collected separately from the ether layer. The combined aq. layers are then neutralised slowly by stirring with aq. NaOH in drops, till the solution becomes alkaline to litmus. The component regenerated is filtered (if it is a solid), or extracted with ether (if it is a liquid), followed by the evaporation of ether.

$$RNH_2 + HCl \rightarrow RNH_3^+ Cl$$

$$RNH_3^+ Cl + NaOH \rightarrow \underset{\text{regenerated}}{RNH_2} + NaCl + H_2O$$

The ether layer from the separating funnel, on evaporation, gives component-II.

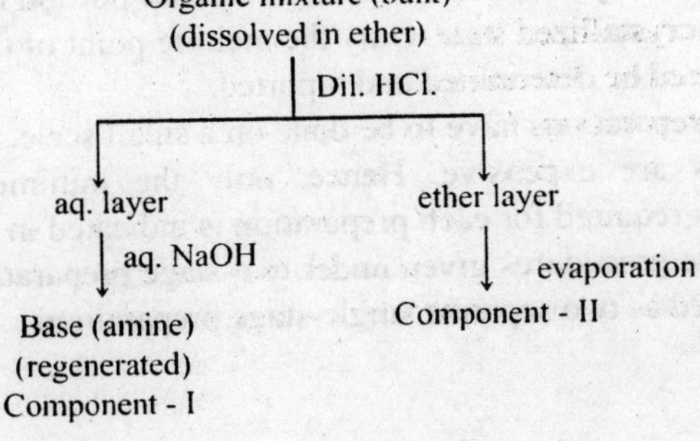

PART B
ORGANIC PREPARATIONS

INTRODUCTION

Different types of reactions are involved in the preparation of an organic compound in the laboratory. Some of the important types of reaction are given below :—

(1) Hydrolysis.	(6) Oxidation.
(2) Acetylation.	(7) Esterification.
(3) Bromination.	(8) Sulphonation.
(4) Nitration.	(9) Reduction.
(5) Benzoylation.	(10) Rearrangement.

Conditions like temperature, relative proportions, concentrations of the reagents, duration of heating, etc., have to be carefully maintained in order to obtain the maximum yield in a pure state. In the following pages, the preparation of some simple organic compounds are described to illustrate the various types of reactions involved.

The preparation of organic compounds, sometimes, involves more than two stages. However, in this part, only organic preparations involving a single stage and two stages are given. In carrying out these experiments, the aim of the student should be to obtain the maximum yield of the product in high state of purity. The student will gradually acquire the necessary skills to set up the apparatus required for refluxing, and to do recrystallisation, etc., neatly and safely.

The students will be expected to submit for evaluation the crude product in a dry condition, as well as a small portion of the product in a pure, recrystallized state. Only the melting point of the recrystallized sample need be determined and reported.

The preparations have to be done on a small scale, since laboratory chemicals are expensive. Hence, only the minimum quantity of chemicals required for each preparation is indicated in the procedures.

Note: The procedures given under two-stage preparations can also be used as two separate single-stage preparations.

B-I SINGLE-STAGE PREPARATIONS

A. HYDROLYSIS

Aromatic esters and amides can easily be hydrolysed by boiling with aqueous NaOH solution. The products can be precipitated on acidification with con. HCl.

(01) Benzoic Acid from Ethyl Benzoate

Principle

Ethyl benzoate is hydrolysed to sodium benzoate and ethanol when heated with aqueous NaOH solution. From sodium benzoate, benzoic acid is liberated by the addition of a mineral acid like HCl.

$$\underset{\text{Ethyl benzoate}}{C_6H_5COOC_2H_5} + \underset{\text{(aqueous)}}{NaOH} \rightarrow \underset{\text{Sodium benzoate}}{C_6H_5COO^- Na^+} + C_2H_5OH$$

$$\underset{\text{Sodiumbenzoate}}{C_6H_5COO^- Na^+} + \underset{\text{(con.)}}{HCl} \rightarrow \underset{\text{Benzoic acid}}{C_6H_5COOH} \downarrow + NaCl$$

Mechanism

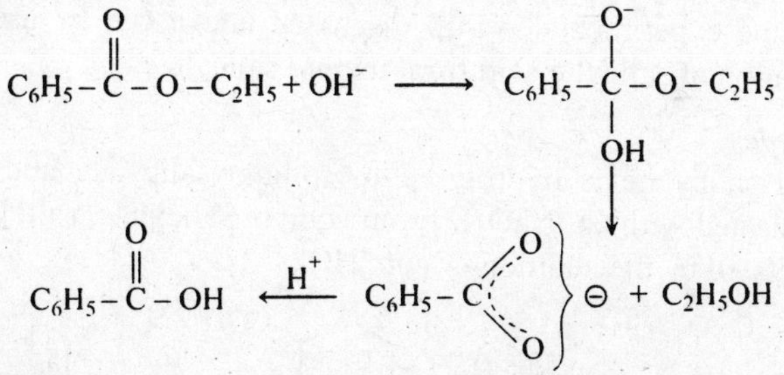

Requirements

Chemicals		***Apparatus***	
Ethyl benzoate	4 ml	Round-bottom flask (250 ml)	1
Aq. NaOH (10%)	25 ml	Reflux condenser	1
Con. HCl	20 ml	Glass beaker (250 ml)	1
Litmus paper (blue)	1	Glass rod	1
		Wire-gauze	1

Procedure

Ethl benzoate (4 ml) and 10 % aq. NaOH (25 ml) are mixed in a round-bottom flask. A bit of broken porcelain is added to prevent bumping. The round-bottom flask is fitted with a reflux condenser, and heated over a wire-gauze steadily till the oily layer of ester disappears (about 30-

45 minutes). Then, the contents of the flask are cooled and transferred into a 250 ml beaker. The flask is rinsed with about 5 ml water and the rinsings are also added to the beaker. To free the benzoic acid from the salt solution, con HCl is added, drop by drop, with constant stirring, till the solution is distinctly acidic (test with litmus paper). Benzoic acid is liberated in the solid form. The contents of the beaker are cooled, and the precipitated benzoic acid is filtered off at the pump (water pump or vacuum pump), washed with cold water, dried using filter paper and the yield of the dry crude product in noted.

(Expected yield : 3 g.)

Recrystallisation

About 0.5 g of the crude benzoic acid is dissolved in 5 ml of boiling water, filtered, poured into a watch glass and allowed to cool slowly. After the crystals are formed the liquid is poured off and the crystals are dried with bits of filter paper. The melting point is determined.

(Correct melting point : 121°C.)

(02) Salicylic Acid from Methyl Salicylate

The same experimental set-up discussed above can be used for the preparation of salicylic acid from methyl salicylate.

Principle

Methyl salicylate is hydrolysed to sodium salicylate and methanol when heated with aq. NaOH. From sodium salicylate, salicylic acid is precipitated by the addition of con. HCl.

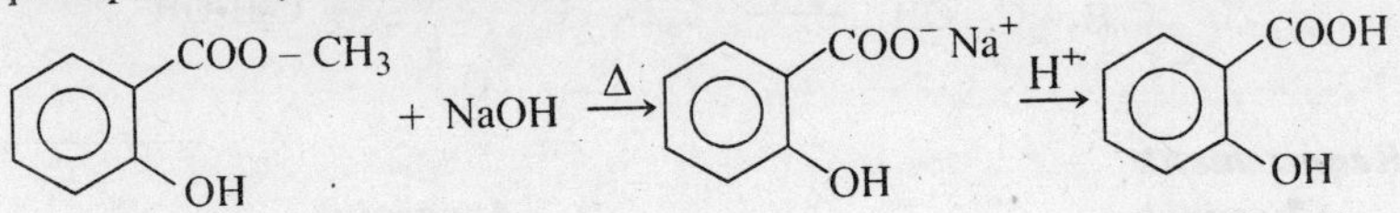

Methyl salicylate | Salicylic acid

Requirements

Methyl salicylate	4 ml
Aq. NaOH soln. (10 %)	30 ml
Con. HCl	20 ml

Apparatus : As in preparation 01.

Procedure

The procedure to be adopted is the same as that given for preparing benzoic acid from ethyl benzoate (see above).

(Expected yield : 4 g)

Recrystallisation

About 0.5 g of the crude salicylic acid is recrystallised from 5 ml boiling water and the melting point is determined.

(Correct melting point : 159° C.)

(03) Benzoic Acid from Benzamide

Principle

An aromatic amide can easily be hydrolysed by a concentrated base or acid into the corresponding acid. Under alkaline conditions, the carboxylic acid is obtained as its salt, from which it can be liberated by the addition of a mineral acid.

$$\underset{\text{Benzamide}}{C_6H_5CONH_2} + NaOH \longrightarrow C_6H_5COO^-Na^+ + NH_3\uparrow$$

$$C_6H_5COO^-Na^+ + HCl \longrightarrow C_6H_5COOH\downarrow + NaCl$$

Requirements

Benzamide	3g.
Aq. NaOH soln. (20 %)	20 ml

Apparatus : As in preparation 01.

Procedure

The experimental set-up and procedure discussed for the preparation of benzoic acid from ethyl benzoate can be followed for this also. During refluxing, ammonia gas is liberated and the contents are heated till no more ammonia comes out from the condenser (about 30–45 minutes). About 0.5 g of the crude product is recrystallised from 5 ml of boiling water. The melting point is determined.

(Expected yield : 3 g. Correct melting point : 121°C.)

Note: Benzamide (3 g) can also be hydrolysed in acid medium using 30 ml con. HCl. After heating the mixture for about 30 minutes, the contents are cooled. Benzoic acid separates out as colourless crystals.

$$\underset{\text{Benzamide}}{C_6H_5CONH_2} + HCl + H_2O \longrightarrow \underset{\text{Benzoic acid}}{C_6H_5COOH\downarrow} + NH_4Cl$$

B. ACETYLATION

Alcohols, phenols and amines (primary and secondary) can easily be acetylated to solid acetyl derivatives, using a mixture of acetic anhydride and glacial acetic acid (or, acetic anhydride and sodium acetate). Acetyl chloride can also be used as an acetylating agent.

Acetyl derivatives of aromatic amines may be prepared either with acetic anhydride or acetic acid or with a mixture of both the reagents. If the heating is prolonged and an excess of acetic anhydride is used, then diacetyl derivatives will also be formed. However, the diacetyl derivatives are unstable in the presence of water, undergoing hydrolysis to the mono-acetyl derivatives. So, when they are crystallised from an aqueous solvent, only mono-acetyl derivative is obtained.

(01) Acetanilide from Aniline

Principle

Aniline can be acetylated easily by refluxing with a mixture of acetic anhydride and sodium acetate.

$$\underset{\text{Aniline}}{C_6H_5NH_2} + (CH_3CO)_2O \xrightarrow{CH_3COO^- \, Na^+} \underset{\text{Acetanilide}}{C_6H_5NHCOCH_3} + CH_3COOH.$$

Mechanism

Aniline acts as a nucleophile. The lone pair of electrons present on the nitrogen atom attacks the carbonyl group of acetic anhydride, and the resulting product is acetanilide.

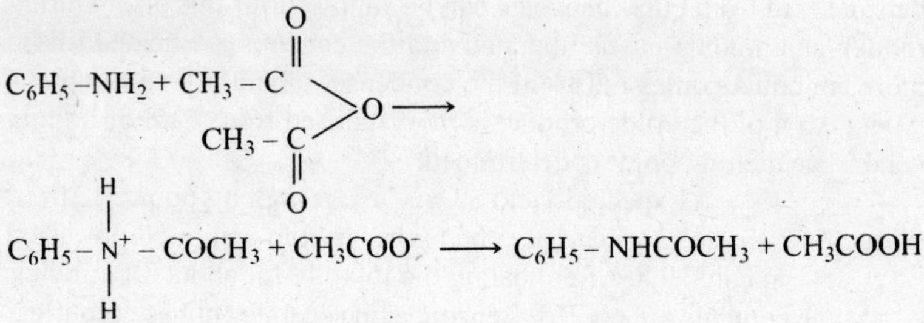

Requirements

Chemicals		*Apparatus*	
Aniline freshly distilled	3 ml	Round bottom flash (250ml)	1
Acetic anhydride	4 ml	Reflux condenser	1
Glacial acetic acid	4 ml	Wire-gauze	1
		Glass beaker (250 ml)	1
		Test tube (10 ml)	1

Procedure

The glacial acetic acid is mixed with acetic anhydride in a dry test tube and shaken well. Aniline is taken in a round bottom flask and the above mixture is added slowly to aniline with constant shaking. A bit of broken porcelain is added to prevent bumping. The flask is fitted with a reflux condenser, and heated gently for about 30 minutes over a wire-gauze. The hot liquid is then poured in a thin stream, with stirring, on crushed ice in a beaker. The white solid that separates is filtered, washed with cold water, dried with filter paper, and the yield is noted.

(Expected yield : 4.5 g.)

Recrystallisation

About 1 g of the crude acetanilide sample is dissolved in the minimum amount hot water, 1 g of animal charcoal is added and then heated. The hot liquid is rapidly filtered (through hot water funnel, if available). The filtrate is allowed to cool. The acetanilide crystalises out in rhombic plates.

(Correct melting point : 114°C)

(02) Acetyl Salicylic Acid (Aspirin) from Salicylic Acid

Principle

Salicylic acid is a phenolic acid. The phenolic group can easily be acetylated using acetic anhydride or acetyl chloride. When salicylic acid is boiled with acetic anhydride, acetyl salicylic acid is obtained.

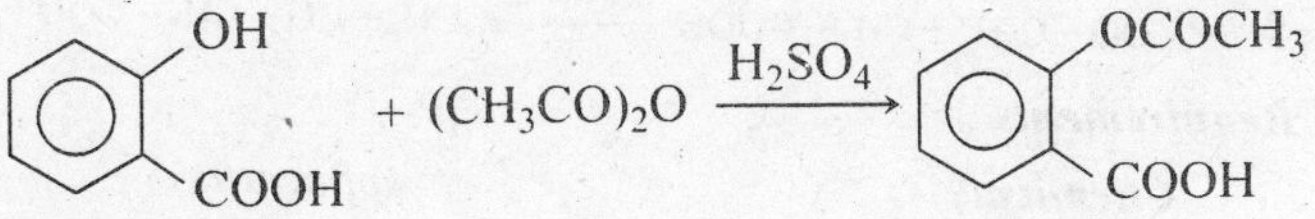

Salicylic acid   Aspirin

Requirements

Chemicals.		*Apparatus.*	
Salicylic acid	3 g.	Conical flask (100 ml)	1
Acetic anhydride	5 ml	Water-bath	1
Con. sulphuric acid	3	Glass beaker (500 ml)	1
Ethanol (rectified spirit)	10 ml		

Procedure

Salicylic acid (3g) is mixed with acetic anhydride (5 ml) in a 100 ml conical flask, and 3 drops con. sulphuric acid is added. The mixture is shaken well, and heated gently on a water-bath at 50–60°C for abo

10–15 minutes, with occasional shaking. Then the conical flask is removed, the contents are cooled under the tap, and then poured in a thin stream into 100 ml of water in a beaker. The acetyl salicylic acid that separates out is filtered off at the pump, washed with water, dried with filter paper and weighed.

(Expected yield : 4 g.)

Recrystallisation

The crude acetyl salicylic acid is recrystallised as follows : About 1 g of the sample is dissolved in the minimum amount of hot ethanol and then poured into 40 ml of water in a 250 ml beaker. The turbid solution is then warmed to get a clear solution. The resulting clear solution is filtered and allowed to cool slowly, when needles of acetyl salicylic acid separate out. It does not possess a definite melting point (decomposes at 128–135°C.)

Note: Acetyl salicylic acid can also be recrystallised from a mixture of equal volumes of acetic acid and water.

(03) ETHYL ACETOACETATE FROM ETHYL ACETATE

Principle

Ethyl acetate consists of an active methylene group which reacts with glacial acetic acid to form ethyl acetate in the presence of sodium metal.

$$CH_3{-}COO{-}C_2H_5 + CH_3{-}COOH \xrightarrow{C_2H_5ONa} CH_3{-}CO{-}CH_2{-}COO{-}C_2H_5 + H_2O$$

Requirements

Chemicals		*Apparatus*	
Ethyl acetate	75 g	Round-bottom flask 250 ml	1
Sodium (metal)	7.5 g	Leibig condenser	1
Glacial acetic acid	25 g	Separating funnel	1

Procedure

The apparatus used in this preparation should be washed well and dried perfectly and protected from moisture by a calcium chloride tube.

Dry metal sodium is taken in a dry 500 ml round-bottom flask fitted with a reflux condenser attached to a calcium chloride guard tube. The flask is surrounded by cold water and the ethyl acetate is added all at once through the condenser. The guard tube is then replaced after the addition is over. The flask is heated on a water bath till all the sodium has disappeared. It may take more than one hour. The warm liquid is treated gradually with glacial acetic acid diluted with water till it is just acid to litmus.

The neutralised liquid is transferred to a separating funnel, an equal volume of saturated sodium chloride solution is added and the mixture is shaken. The upper layer consisting of a mixture of ethyl acetate and ethyl acetoacetate is separated, dried and distilled on a boiling water bath to remove ethyl acetate.

It is a colourless liquid possessing fruity smell.

(Expected yield : 10 g, boiling point : 88°C)

(04) PHENACETIN FROM P-AMINOPHENOL

Principle

Phenacetin is p-ethoxyacetanilide. It is normally prepared from p-aminophenol by first acetylation into p-hydroxy acetanilide and then by etherification into p-ethoxyacetanilide in the presence of sodium ethoxide.

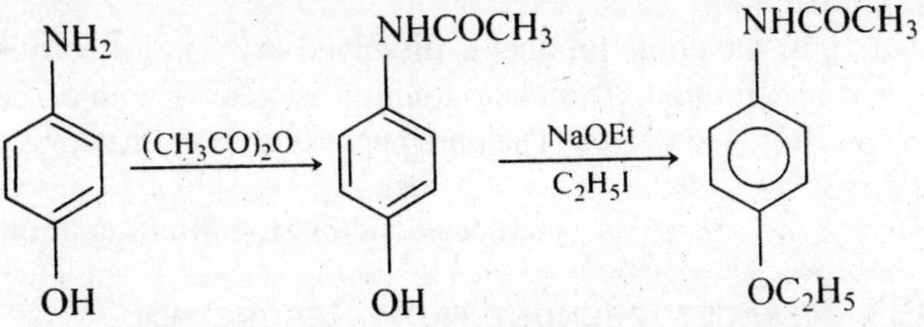

p - amino phenol

Phenacetin

Requirements

Chemicals		***Apparatus***	
p-Aminophenol	10 g	Round bottom flask 250 ml	1
Acetic anhydride	12 g	Beaker 250 ml	1
Sodium metal	2 g	Test tubes	3
Absolute alcohol	40 ml	Water bath, Funnel	1
Ethyl iodide	8 ml	Watch glass	1

Procedure

The p-aminophenol are suspended in 30 ml of water contained in a 250 ml beaker or conical flask. About 12 ml of acetic anhydride are added with vigorous stirring. The reaction mixture is then warmed on a water bath. The solid dissolves. After 10 minutes, the solid acetyl derivative is filtered after cooling, washed with a little cold water, recrystallised from hot water and dried upon filter paper in the air. The yield of p-acetylaminophenol is noted.

(Expected yield: 14g; Melting point : 169°C)

The absolute alcohol is taken in a 250 ml round-bottom flask fitted with a reflux condenser, and metal sodium is added. Vigorous reaction starts. If the reaction is slow, then the flask is warmed on a water bath until the reaction is complete. The mixture is cooled and p-acetylaminophenol are added. The ethyl iodide are added slowly through the condenser and the mixture is refluxed for about 45 minutes. 100 ml of water are then added through the condenser if crystals do separate, the mixture is refluxed until the crystals redissolve.

The flask is then cooled in an ice bath. The crude phenacetin which separates is filtered and washed with little cold water and dried.

Recrystallisation

About 1g of the crude product is dissolved in 15 ml of rectified spirit, refluxed and filtered. The clear solution is treated with 50 ml of hot water and allowed to cool. The pure phenacetin is collected by filtration and dried.

(Expected yield: 10 g; Melting point: 136°C)

(05) RESACETOPHENONE FROM RESORCINOL

Principle

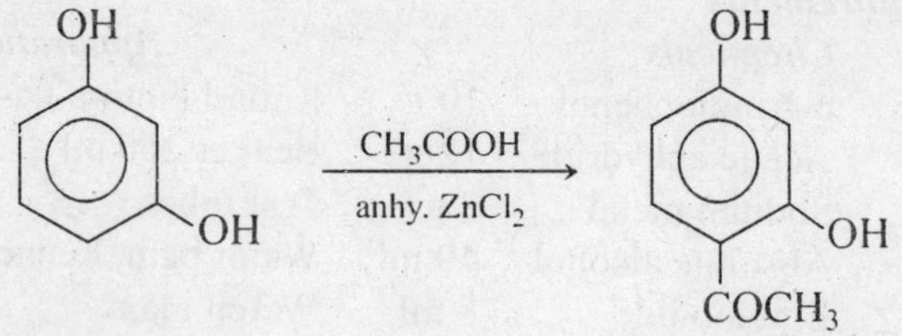

Resorcinol Resacetophenone

Requirements

Chemicals		*Apparatus*
Zinc chloride (anhydrous)	15 g	Porcelain basin
Glacial acetic acid	15 g	Desiccator
Resorcinol	10 g	Beaker 100 ml,
Con. HCl	40 ml	Water bath, glass rod, funnel

Procedure

The zinc chloride is fused in a porcelain basin until it shows a tendency to fume, then poured into a clean mortar, powdered quickly and kept in a desiccator, ready for use. Glacial acetic acid (frozen and remelted) is taken in a beaker and the zinc chloride dissolved in it by heating. Resorcinol is added slowly with constant stirring. The mixture is heated just to boiling (142°C) and cooled. The syrupy liquid is then treated with a mixture of concentrated hydrochloric acid and water(1:1). A crystalline mass is obtained slowly. The crystals are filtered off, washed with water and dried.

Recrystallisation

A small portion of the sample is recrystallised from boiling water. Pale orange yellow plates are obtained.

(Expected yield: 8g; Melting point: 143°C)

C. BROMINATION

Benzene and alkyl substituted benzenes do not react appreciably with chlorine and bromine in the cold, but in the presence of halogen carriers such as Al / Hg, I_2/Fe, reaction takes place readily. The reaction must be carried out in the absence of direct sunlight, since sunlight causes direct addition of the halogen to the hydrocarbon.

Aromatic compounds containing activating groups like –OH, $-NH_2$ and $-NHCOCH_3$ can easily form bromoderivatives, when treated with a solution of bromine in acetic acid. Aniline and phenol can form a tribromo derivative with excess bromine in acetic acid medium. The solid separates out on dilution with water.

Note: Liquid bromine is highly corrosive, and must be handled very carefully. A solution of bromine in glacial acetic acid can be prepared in bulk, and the required amount can be given to each student.

(01) 2,4,6-TRIBROMO ANILINE FROM ANILINE

Principle

When treated with an excess of bromine in glacial acetic acid, aniline gives symmetrical tribromoaniline.

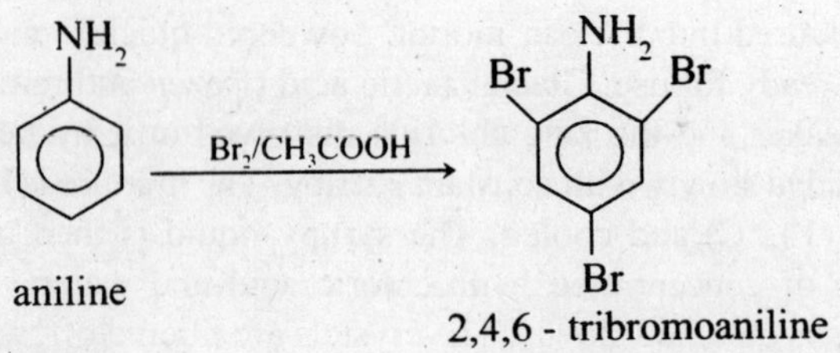

Requirements

Chemicals		*Apparatus*	
Aniline (distilled)	2 ml	Conical flask (250 ml)	1
Bromine	3 ml	Cork	1
Glacial acetic acid	15 ml	Boiling test tube	1
Rectified spirit	5 ml	Dropper	1

Procedure

Aniline is dissolved in about 5 ml of glacial acetic acid in a conical flask fitted with a cork. A solution of bromine in glacial acetic acid (3 ml bromine in 10 ml glacial HOAc) is taken in a boiling test tube. Using a dropper, about 1 ml of bromine in acetic acid is added at a time to the contents of the flask, and shaken well. The addition of bromine is continued till the colour of the solution in the flask becomes yellow. The flask is corked well and shaken vigorously for about 10–15 minutes. About 100ml of cold water is then added to the conical flask, and the mixture is shaken vigorously for two minutes. Colourless crystals of tribromoaniline that separate out are filtered off at the pump, washed with water, dried and weighed.

(Expected yield : 6 g.)

Recrystallisation

About 1 g of the crude sample is recrystallised from hot 1 : 1 aqueous alcohol, and the melting point is determined.

(Correct melting poiunt : 120ºC.)

(02) p-BROMOACETANILIDE FROM ACETANILIDE

Principle

The anilide group present in acetanilide is a moderate activating group, which directs the incoming bromonium ion to *ortho* and *para* positions. Practically, only the *para*-product is formed, due to the steric hindrance of the bulky functional group.

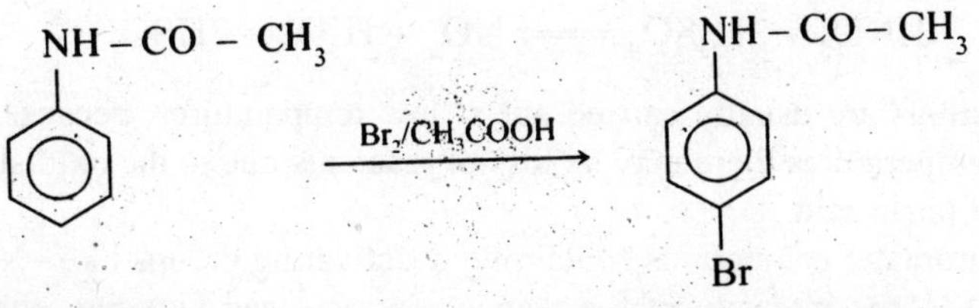

acetanilide p - bromoacetanilide

Requirements

Chemicals		Apparatus	
Acetanilide	2 g.	Conical flask (250 ml)	1
Bromine	3 ml	Cork	1
Glacial acetic acid	15 ml	Boiling test tube	1
Rectified spirit	10 ml	Dropper	1
		Glass rod	1

Procedure

Acetanilide (2 g) is dissolved in 5 ml glacial acetic acid in a conical flask, fitted with a good cork. Bromine (3 ml) is dissolved in 10 ml glacial acetic acid and taken in a boiling test tube. Using a dropper, about 1 ml of bromine in acetic acid is added at a time to the contents of the flask, and shaken well. The addition of bromine is continued till the colur of the solution in the flask becomes yellow. The flask is corked well and shaken vigorously for about 10–15 minutes. The flask is kept aside for 10 minutes. The contents of the flask are then poured in a thin stream into crushed ice in a beaker, with stirring. The colourless crystals of *p*-bromoacetanilide that separate out are filtered, washed and dried.

(Expected yield: 2.5 g)

Recrystallisation

About 0.5 g of crude acetanilide is recrystallised from hot rectified spirit, and the melting point is determined.

(Correct melting point : 167ºC)

D. NITRATION

Aromatic hydrocarbons may be nitrated with con. nitric acid, in the presence of con. sulphuric acid. Nitration of an aromatic compound is an example of aromatic electrophilic substitution. The function of the sulphuric acid is to furnish a strongly acid medium, and to convert the nitric acid into the highly reactive nitronium ion, NO_2^+, which is the real nitrating agent, acting as the electrophile.

$$HNO_3 + 2H_2SO_4 \rightleftharpoons NO_2^+ + H_3O^+ + 2HSO_4^-$$

Nitrations are usually carried out at low temperatures, because at higher temperatures there may be loss of reactants due to the oxidising action of nitric acid.

For aromatic compounds containing deactivating groups like $-NO_2$ and $-SO_3H$, the intensity of the reaction is increased by using either fuming sulphuric or fuming nitric acid.

(01) m-DINITROBENZENE FROM NITROBENZENE

Principle

Nitrobenzene, on nitration using a mixture of con. sulphuric acid and fuming nitric acid, gives m-dinitrobenzene.

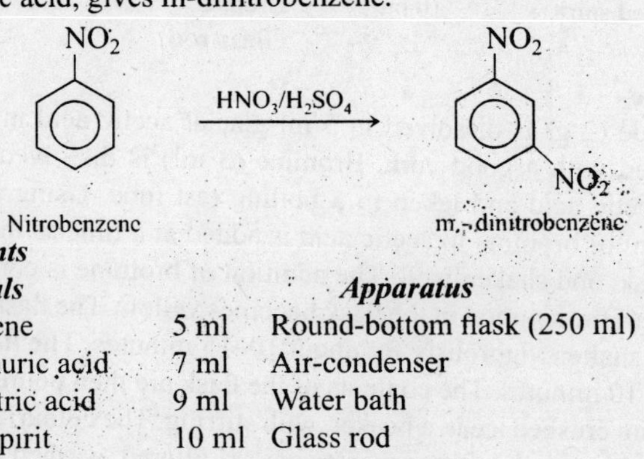

Nitrobenzene → m-dinitrobenzene

Requirements

Chemicals		*Apparatus*	
Nitrobenzene	5 ml	Round-bottom flask (250 ml)	1
Con. sulphuric acid	7 ml	Air-condenser	1
Fuming nitric acid	9 ml	Water bath	1
Rectified spirit	10 ml	Glass rod	1
		Beaker (250 ml)	1

Procedure

Fuming nitric acid (9 ml) is first taken in a round-bottom flask, and con. sulphuric acid (7 ml) is added slowly with stirring. The round-bottom flask is cooled in the tap water during the addition. Then, nitrobenzene (5 ml) is added, in small quantities at a time, and the reaction mixture is shaken vigorously after each addition. The round-bottom flask is fitted with air-condenser, and the reaction mixture is heated in a boiling water-bath for about an hour. The completion of reaction is tested by adding a few drops of the reaction mixture to water. If the reaction is complete, dinitrobenzene separates out as a solid. Otherwise, the heating is continued for some more time. The contents of the flask are then poured, in a thin stream, into about 100 ml of water in a 250 ml beaker, with vigorous stirring. The yellow solid (*m*-dinitrobenzene) that separates out is filtered off at the pump, washed with cold water thoroughly to remove all the acid, dried with filter paper, and the yield is noted.

(Expected yield : 5g.)

Recrystallisation

About 1 g of the crude m-dinitrobenzene is recrystallised from hot rectified spirit.

(Correct melting point : 90°C)

(02) PICRIC ACID FROM PHENOL

Principle

Phenol on nitration, using the nitration mixture of con. HNO_3 and con. H_2SO_4, gives picric acid (2,4,6–trinitrophenol).

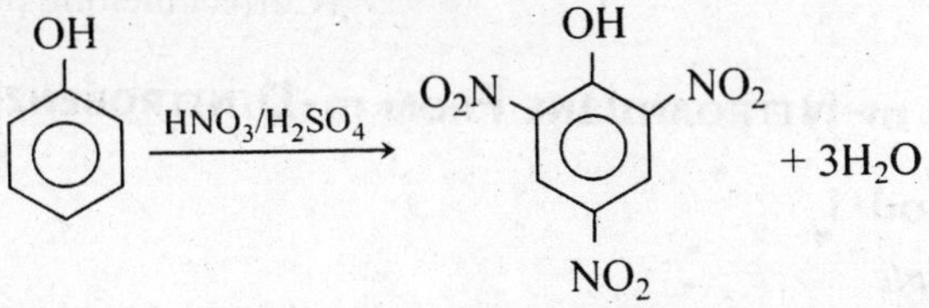

Phenol

Picric acid (or)
2,4,6 - trinitrophenol

Requirements

Chemicals		***Apparatus***	
Phenol (distilled)	3 ml	Round-bottom flask (250 ml)	1
Con. nitric acid	12 ml	Air condenser	1
Con. sulphuric acid	4 ml	Water - bath	1
Rectified spirit	10 ml	China dish	1
		Glass rod	1
		Dropper	1

Procedure

Con. sulphuric acid (4 ml) is taken in a China dish and phenol (3 ml) is added slowly with stirring. The reaction mixture is homogenised by warming on a water-bath. It is then cooled well. Phenol sulfonic acid [C_6H_4 (OH) SO_3H] is formed.

Con. nitric acid (12 ml) is taken in a 250 ml round - bottom flask and cooled well. The phenol sulphonic acid is added in drops, with constant shaking, to the nitric acid. During this mixing, heat is liberated. So, the mixture is cooled well under the tap now and then during the addition, so that the temperature does not rise above 60ºC. Red brown fumes are evoled. When the reaction has subsided, the contents are heated on a boiling water-bath for two hours. The flask is cooled, and about 100 ml cold water is added and the flask is kept in ice taken in a suitable container. Picric acid separates out as a yellow solid. The yellow crystals are filtered off, washed with water thoroughly to remove all acids, dried with filter paper and the yield is noted.

(Expected yield : 5 g.)

Recrystallisation

About 1 g of the crude picric acid is recrystallised from hot rectified spirit, and the melting point is determined.

(Correct melting point : 122ºC.)

(03) m-NITROANILINE FROM m-DINITROBENZENE

Method-1

Principle

The reaction of conversion of m-dinitrobenzene into m-nitroaniline is an example of partial reduction in which one of the nitro group is converted into amino group.

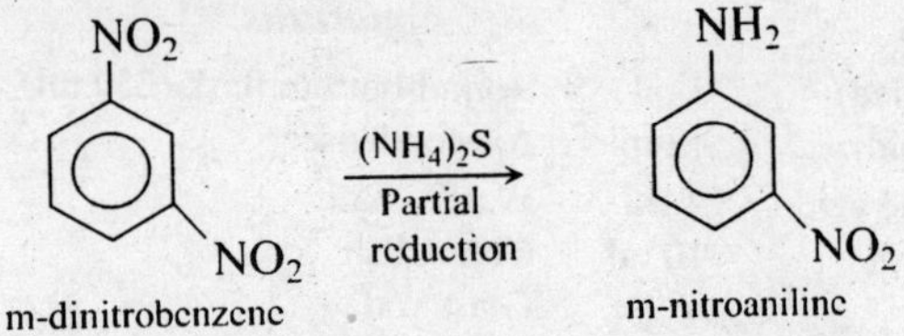

Requirements

Chemicals		***Apparatus***	
m-Dinitrobenzene	10 g	Conical flask (250 ml)	–1
Alcohol (rectified spirit)	40 ml	Water bath	–1
Ammonia (concentrated)	8 ml	Test tube	–3
Hydrogen sulphide solution	20 ml	Glass rod	–1

Procedure

The dinitrobenzene is powdered and taken into a 100 ml conical flask, and dissolved in hot alcohol. The solution is cooled and the ammonia added. The flask is stoppered and a tube is inserted to pass hydrogen sulphide gas. The mixture is saturated with hydrogen sulphide while cooling by immersion in cold water for about an hour. The hydrogen sulphide is washed by passing the gas through a wash bottle containing water. The flask is then heated under reflux on a water-bath for about 15 minutes, cooled to room temperature and hydrogen sulphide passed once again. This process is repeated atleast for 4 times. The liquid is now poured into about 100 ml cold water. The precipitate which is a mixture of m-nitroaniline and sulphur is filtered and washed with cold water. The product is extracted thrice by warming with hot dilute hydrochloric acid (10 to15 ml) in a flask and the extract filtered. A few drops of the filtrate are treated in a test tube with ammonia to see if the extraction is complete. The combined filtrate is concentrated to half the bulk and treated with a slight excess of concentrated ammonia to precipitate completely the m-nitroaniline. The m-nitroaniline is filtered, washed and dried on a water bath.

(Expected yield: 5 g)

Recrystallisation

About 1 g of the crude sample is recrystallised from minimum amount of boiling water. It separates in fine yellow needles.

(Melting point: 112°C)

Method-2

Requirements

Chemicals		*Apparatus*	
m-Dinitrobenzene	10g	Conical flask (250 ml)	1
Sodium sulphide $Na_2S.9H_2O$	2.5g	Beaker (250 ml)	1
Flowers of sulphur	5g	Water bath	1
Sodium hydroxide	10g	Test tube	1
		Glass rod	1

Procedure

Preparation of sodium sulphide: If sodium sulphide is not available, it can be prepared by dssolving 8 g of sodium hydroxide in 30 ml of water, dividing the solution into two equal parts, saturating one portion with hydrogen sulphide and adding the remainder of the alkali. The flowers of sulphur are dissolved in the sodium sulphide.

The m-dinitrobenzene are powdered and suspended in 50 ml of water contain in a 250 ml beaker. The beaker is heated on a water bath. The contents are stirred and the sodium sulphide solution with the sulphur dissolved in it, added from a dropping funnel in the course of about ten minutes. To test whether the reaction is complete, a drop of the solution is placed on a filter paper and touched with ferrous or copper sulphate solution. If the black stain persists for a few seconds, the reduction is over and the mixture is cooled to precipitate the nitroaniline which is purified.

(Expected yield: 5 g)

Recrystallisation

About 1 g of the crude sample is recrystallised from minimum amount of boiling water. It separates in fine yellow needles.

(Melting point: 112°C)

E. SULPHONATION

Con. sulphuric acid is generally employed as a sulphonating agent. When aniline is mixed with an excess of con. H_2SO_4, anilinium sulphate is formed. The exces H_2SO_4 acid is removed by aq. NaOH solution, and the product heated to about 180°C, when *p*-aminobenzene sulphonic acid is formed.

Sulphanilic Acid from Aniline

Principle

When aniline is heated to 180°C with con. sulphuric acid, *p*-aminobenzene sulphonic acid (sulphanilic acid) is formed.

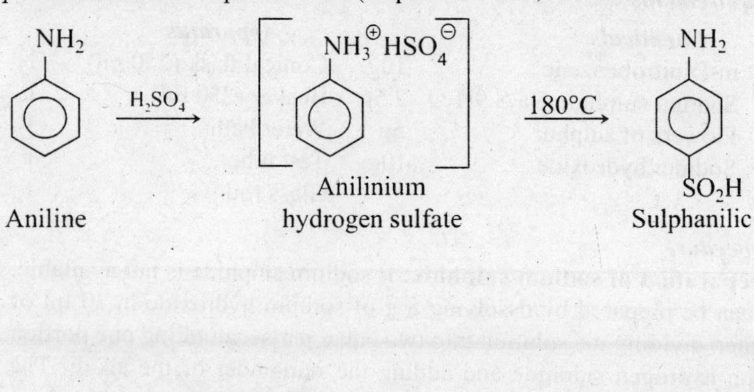

Aniline — Anilinium hydrogen sulfate — Sulphanilic acid

Requirements

Chemicals		Apparatus	
Aniline	3 ml	Round-bottom flask (250 ml)	1
Con. sulphuric acid	6 ml	Air-condenser	1
Aq. NaOH (10 %)	5 ml	Beaker (250 ml)	1
		Oil bath (silicone oil or coconut oil)	1

Procedure

The aniline is taken in a round-bottom flask and con. sulphuric acid is added, drop by drop, with constant shaking. The flask is fitted with an air-condenser and heated in an oil bath for about 4–5 hours at 180–190°C. Two drops of reaction mixture are poured into 5 ml aq. NaOH solution; if not oily drops separate, the reaction is assumed to be complete. After the reaction is over, the contents of the flask are cooled and poured into cold water in a 250 ml beaker with constant stirring. The crude product that separates out is filtered off at the pump, washed with water, dried using filter paper and the yield is noted.

(Expected yield : 5 g.)

Recrystallisation

About 1 g of crude sample is recrystallised from minimum amount of boiling water.

(Melting point : Not sharp)

F. BENZOYLATION

Aromatic amines and phenols usually undergo benzoylation to the corresponding benzoyl derivatives. Benzoyl chloride is the reagent commonly used. (Benzoyl chloride is only very slowly hydrolysed by water, and hence the benzoylation reaction can be carried out in aqueous medium.) Benzoylation proceeds smoothly, and the sparingly soluble benzoyl derivative usually separates as a solid. Generally aq. NaOH solution is used to remove HCl, which also hydrolyses the excess of benzoyl chloride giving sodium benzoate and NaCl, which remain in solution.

$$C_6H_5COCl + 2\ NaOH \rightarrow C_6H_5COO\ \ Na^+ + NaCl + H_2O.$$

In the Schotten-Baumann method of benzoylation, the amino compound is dissolved in an excess of aq. NaOH solution: an excess of benzoyl chloride is added and the mixture is vigorously shaken in a stoppered vessel.

***Note*:** Benzoyl chloride must be used very carefully, since it is lachrymatory (tear producing) reagent.

(01) Benzanilide from Aniline

(Schotten - Baumann Reaction)

Principle

Aniline undergoes benzoylation when shaken well with benzoyl chloride and aq. NaOH solution, giving benzanilide.

$$C_6H_5 - NH_2 + C_6H_5 - CO - Cl \xrightarrow{NaOH} C_6H_5NHCOC_6H_5 + HCl$$

Requirements

Chemicals		Apparatus	
Aniline	3 ml	Conical flask (250 ml)	1
Benzoyl chloride	5 ml	Rubber cork	1
Aq. NaOH (10 %)	40 ml	Test tube	1
Methylated spirit	10 ml	Dropper	1

Procedure

Aniline (3 ml) and NaOH (40 ml) are mixed in a 250 ml conical flask, fitted with a good rubber cork and shaken well.

The benzoyl chloride is added slowly, 1 ml at a time, and shaken well. After the addition of benzoyl chloride is over, the conical flask is corked tightly and shaken vigorously for about 15–20 minutes. The disappearence of the smell of benzoyl chloride and the separation of a brown solid indicates the completion of reaction. If the smell of benzoyl chloride still persits, add 5 ml more of aq. NaOH and shake well for 2–3 minutes. When the reaction is complete, the flask is cooled well and the product is filtered off at the pump, washed well with water, dried with filter paper and the yield is noted.

(Expected yield : 6g)

Recrystalisation

About 0.5 g of crude benzanilide is recrystallised from hot methylated spirit.

(Correct melting point : 162°C.)

(02) 2-Naphthyl Benzoate from 2-Naphthol

Principle

The procedure to be adopted is the same as that given above for benzanilide (Preparation 01).

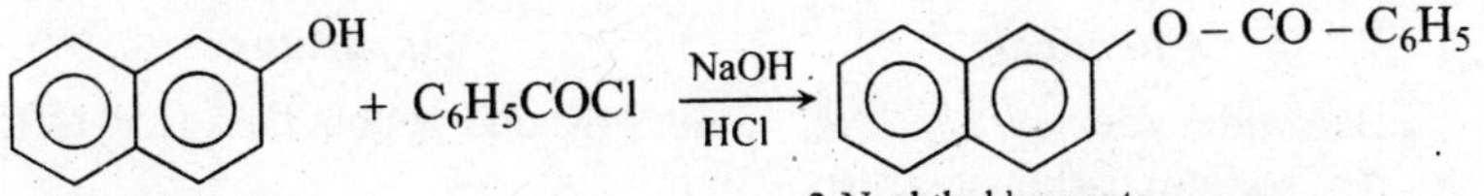

2-Naphthol 2-Naphthyl benzoate

(Expected yield : 4 g)

Recrystallisation

About 0.5 g of crude 2-naphthyl benzoate is recrystallised from hot methylated spirit.

(Correct melting point : 107°C)

G. OXIDATION

When an aromatic compound containing a –CHO or —CH_2OH group is oxidised using alkaline permanganate solution, benzoic acid results.

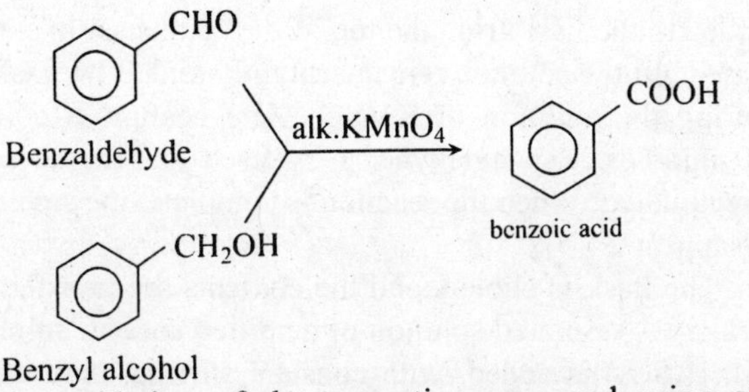

During the oxidation of the aromatic compound using alkaline $KMnO_4$ solution, manganese dioxide gets precipitated. Usually acidified sodium sulphite solution is added to dissolve the precipitated MnO_2. Finally, the carboxylic acid is precipitated on acidification with con. HCl.

(01) Benzoic Acid from Benzaldehyde

Principle

When benzaldehyde is oxidised using alkaline $KMnO_4$ soln. benzoic acid is formed in the form of its salt. The precipitated manganese dioxide is dissolved using acidified sodium sulphite solution. Benzoic acid is then precipitated on acidifying the resulting solution, using con. HCl.

$$\underset{\text{Benzaldehyde}}{C_6H_5CHO} \xrightarrow[\text{(2) con. HCl}]{\text{(1) alk.KMnO}_4} \underset{\text{Benzoic acid}}{C_6H_5COOH \downarrow}$$

Requirements

Chemicals		*Apparatus*	
Benzaldehyde	3 ml	Round-bottom flask (250 ml)	1
Aq. sodium carbonate (10%)	30 ml		
$KMnO_4$	5 g.	Reflux condenser	1
Sodium sulphite	5 g.	Wire-gauze	1
Dil. sulphuric acid	20 ml	Beaker (250 ml)	1
Con. HCl	20 ml	Glass rod	1

Procedure

Benzaldehyde (3 ml) is mixed with aq. sodium carbonate soln. taken in a round-bottom flask. A bit of broken porcelain is added, and the round-bottom flask is fitted with a reflux condenser. The contents are gently boiled by heating over a wire-gauze. An aq. soln. of $KMnO_4$ (prepared by dissolving 5g in the minimum amount of water) is prepared, and added to the flask from the top of the condenser in small quantities at a time until the solution remains slightly pink. The flask is shaken gently during the addition of $KMnO_4$. The heating is continued for about 30 minutes. Benzaldehyde is oxidised to benzoic acid and MnO_2 is precipitated. When the reaction is complete, the smell of benzaldehyde disappears.

The flask is cooled, and the contents are transferred into a 250 ml beaker. A saturated solution of acidified sodium sulphite (5 g in 20 ml dil. H_2SO_4) is added, with constant stirring, until all the precipitated MnO_2 is dissolved. Then, to liberate the free acid, con. HCl is added till precipitation is complete. The precipitated acid is filtered off, washed with water, dried using filter paper and the yield is noted.

(Expected yield : 5g.)

Recrystallisation

About 1 g of the crude benzoic acid is recrystallised from the minimum amount of boiling water.

(Correct melting point. 121°C.)

(02) Benzoic Acid from Benzyl Alcohol

Principle

When benzyl alcohol is oxidised using alkaline $KMnO_4$ soln., benzoic acid is formed as its sodium salt. The precipitated MnO_2 is dissolved using acidified sodium sulphite solution. On acidification, benzoic acid precipitates out.

$$\underset{\text{Benzyl alcohol}}{C_6H_5CH_2OH} \xrightarrow[\text{(2) con. HCl}]{\text{(1) alk.}KMnO_4} \underset{\text{Benzoic acid}}{C_6H_5COOH \downarrow}$$

Requirements

Chemicals		*Apparatus*	
Benzyl alcohol	3 ml	Round-bottom flask (250 ml)	1
Aq. sodium carbonate (10 %)	30 ml		1
$KMnO_4$	4 g.	Reflux condenser	1
Sodium sulphite	15g.	Wire-gauze	1
Dil. sulphuric acid	20 ml	Beaker (250 ml)	1
Con. HCl	20 ml	Glass rod	1

Procedure

The benzyl alcohol is mixed with aq. sodium carbonate soln. and taken in a 250 ml round-bottom flask. A bit of broken porcelain is added, and the round-bottom flask fitted with a reflux condenser. The contents are gently boiled by heating over a wire-gauze. An aq. soln. of $KMnO_4$ is prepared and added to the flask from the top of the condenser, in small quantities at a time, until the solution remains slightly pink. The contents of the flask are refluxed, for about 90 minutes. The round-bottom flask is cooled, and the contents are transferred into a 250 ml beaker. A saturated soln of acidified sodium sulphite (15 g in 20 ml dil. H_2SO_4) is added, with constant stirring, until all the precipitated MnO_2 gets dissolved. To liberate the free acid, con. HCl is added till the precipitation is complete. The precipitated acid is filtered off, washed with water, dried and the yield is noted.

(Expected yield : 3g.)

Recrystallisation

About 1 g of the crude benzoic acid is recrystallised from the minimum amount of boiling water.

(Correct melting point : 121°C.)

(03) Benzoic Acid from Toluene

Principle

Toluene on oxidation using alkaline potassium permanganate gives benzoic acid. When toluene is oxidized using alkaline $KMnO_4$ solution, benzoic acid is formed in the form of its salt. The precipitated manganese dioxide is dissolved using acidified sodium sulphite solution. Benzoic acid is then precipitated on acidifying the resulting solution, using conc. HCl.

$$\underset{\text{Toluene}}{C_6H_5 - CH_3} \xrightarrow{\text{Alkaline } KMnO_4} \underset{\text{Benzoic acid}}{C_6H_5 - COOH}$$

Requirements

Chemicals		*Apparatus*	
Potassium permanganate 6.2 g (in 100 ml water)		Round bottom flask 250 ml	1
		Reflux Condenser	1
Toluene	10 ml	Wire gauze	1
Sodium carbonate(2N)	20 ml	Dropping funnel	1
Saturated SO_2 solution	25 ml	Test tube and glass rod	

Procedure

20 ml of sodium carbonate solution (2N) and 10 ml of the toluene are placed in a 250 ml round-bottom flask, fitted with a reflux condenser. The permanganate solution is added in portions from a dropping funnel to the RB flask. The flask is heated on an asbestos-centered wiregauze. The contents of the flask are kept steadily boiling and the addition of permanganate is continued till no more is taken up. The flask is cooled, the precipitated manganese dioxide is filtered, the filtrate is concentrated and then saturated with sulphur dioxide till the solution is colourless(sulphur dioxide solution may be prepared by acidifying sodium bisulphite). On cooling, benzoic acid crystallize out. It is filtered, washed with cold water and dried.

(Expected Yield: 3g)

Recrystallisation

About 1g of the sample is recrystallised from minimum amount of boiling water.

(Melting point:121°C).

Note: 100 ml of $KMnO_4$ solution containing 6.2g will provide 1 g of available oxygen.

(04) Benzoic acid from Benzyl chloride

Principle

Benzyl chloride on oxidation using alkaline potassium permanganate gives benzoic acid. When benzyl chloride is oxidized using alkaline $KMnO_4$ solution, benzoic acid is formed in the form of its salt. The precipitated manganese dioxide is dissolved using acidified sodium sulphite solution. Benzoic acid is then precipitated on acidifying the resulting solution, using conc. HCl.

$$\underset{\text{Benzyl chloride}}{C_6H_5-CH_2-Cl} \xrightarrow{\text{Alkaline } KMnO_4} \underset{\text{Benzoic acid}}{C_6H_5-COOH}$$

Requirements

Chemicals	***Apparatus***	
Potassium permanganate 6.2 g (in 100 ml water)	Round bottom flask 250 ml	1
	Reflux Condenser	
Benzyl chloride 5 ml	wire gauze	
Potassium carbonate 4g/20 ml water	Dropping funnel	
Saturated SO_2 solution 25 ml	Test tube and glass rod	1

Procedure

20 ml of potassium carbonate solution and 5 ml of the benzyl chloride are placed in a 250 ml round-bottom flask, fitted with a reflux condenser. The permanganate solution is added in portions from a dropping funnel to the RB flask. The flask is heated on an asbestos-centered wiregauze. The contents of the flask are kept steadily boiling and the addition of permanganate is continued till no more is taken up. The flask is cooled, the precipitated manganese dioxide is filtered, the filtrate is concentrated and then saturated with sulphur dioxide till the solution is colourless(sulphur dioxide solution may be prepared by acidifying sodium bisulphite). On cooling, benzoic acid crystallize out. It is filtered, washed with cold water and dried.

(Expected Yield: 4 g)

Recrystallisation

About 1g of the crude sample is recrystallised from minimum amount of boiling water.

(Melting point: 121°C)

(05) TERE-PHTHALIC ACID FROM P-XYLENE

Principle

Any alkyl group attached to the benzene nucleus on oxidation is converted into a carboxylic acid group. p-Xylene on oxidation using acidified sodium dichromate gives tere-phthalic acid in which both the methyl groups attached to the benzene nucleus in its 1 and 4 positions are oxidized to carboxylic acid groups.

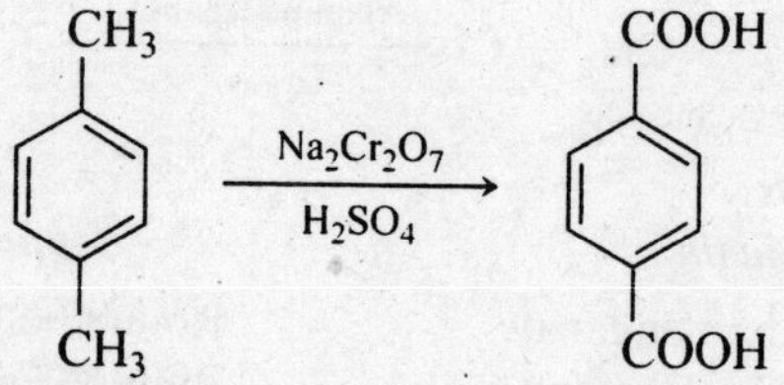

p-Xylene Tere pythalic acid

Method-1

Requirements

Chemicals		*Apparatus*	
p-Xylene	15 ml	Three necked flask (250 ml)	1
Sodium dichromate	70 g/300 ml of water		
Concentrated sulphuric acid	100 ml	Reflux condenser	1
Ether	20 ml	Mechanical stirrer	1
10% NaOH solution	25 ml	Separating funnel	1
		Dropper, test tube, glass rod	

Procedure

In a 1 litre three-necked flask, 15 ml of p-xylene, 70 g of sodium dichromate crystals and 300 ml of water are placed, fitted with a reflux condenser, a mercury-sealed mechanical stirrer and a separating funnel. About 100 ml of concentrated sulphuric acid is added dropwise through the separating funnel to the mixture. The mixture is stirred well at such a rate that the reaction is under control for about 30 minutes. When all the acid has been introduced and the temperature begins to fall, the mixture is refluxed gently for half an hour. The reaction mixture is cooled and poured into ice water in a beaker and allowed to stand for 1 hour.

The crude tere-phthalic acid is filtered, washed with cold water followed by small amount of ether and dried.

(Expected Yield: 9 g)

Recrystallisation

The acid is purified by dissolving in 5% NaOH solution , the solution is filtered from any chromic hydroxide and added to sulphuric acid(20 ml of con. Sulphuric acid in 200 ml of water) The purified acid is filtered and washed thoroughly with cold water and dried at 100°C. Tere-phthalic acid sublimes without melting at 300°C and is almost insoluble in water and alcohol.

Method-2:

Requirements

Chemicals		Apparatus	
Potassium permanganate	25 g	Round-bottom flask 1 litre	1
p-Xylene	10 ml	Leibig condenser	1
Sodium hydroxide (2N)	5 ml	Measuring cylinder	1
Water	250 ml	Beaker 250 ml	1

Procedure

About 20 g of potassium permanganate, 5 ml of NaOH solution and 250 ml of water are taken in a round-bottom flask of 1 litre capacity fitted with a Liebig's condenser. p-xylene is added to the flask finally slowly. A bit of porous porcelain is added to avoid bumping and the contents are refluxed for 3-4 hours. When the colour of the permanganate has disappeared, further quantities of permanganate are added till a faint pink colour persists. The flask is cooled, the manganese dioxide is filtered off, washed with water, the filtrate and washings are concentrated, cooled and saturated with sulphur dioxide. The precipitated acid is filtered and dried.

(Expected Yield 4 g)

(06) p-Nitrobenzoic Acid from p-Nitrotoluene

Principle

p-Nitrotoluene on oxidation gives p-Nitrobenzoic acid.

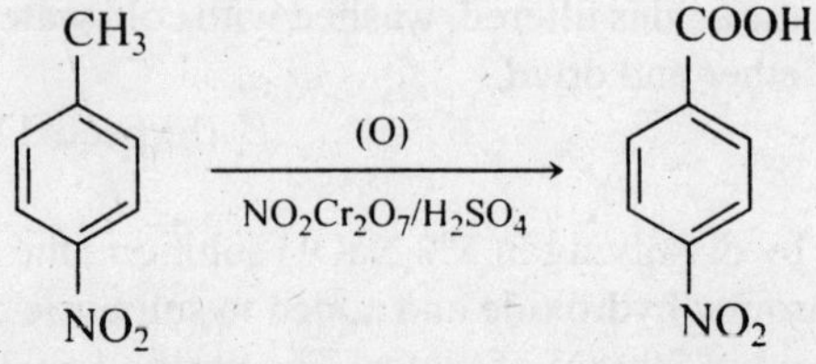

p - nitro toluene p- nitro benzoic acid

Requirements

Chemicals		*Apparatus*
p-Nitrotoluene	4.5 g	Round bottom flask 250 ml
Sodium dichromate	14 g	Beaker 250 ml
Con. Sulphuric acid	20 ml	Funnel
NaOH(5% solution)	40 ml	glass rod
Ethanol	10 ml	wire gauze

Procedure

A mixture of p-nitrotoluene, sodium dichromate crystals and about 30 ml water is taken in a round-bottom flask, fitted with a reflux condenser. Gradually concentrated sulphuric acid is added 1ml at a time with continual stirring. The reaction mixture is heated on a wire-gauze until oxidation begins. The reaction may become vigorous. The reaction mixture is allowed to stand for some time until the vigorous boiling subsides. The reaction mixture is then heated on a gentle flame for about 1½ to 2 hours. The reaction mixture is then cooled and poured into 100 ml water in a 250 ml beaker with constant stirring. The precipitated p-nitrobenzoic acid is filtered, washed with cold water. The crude solid is then washed with 5% aqueous sodium hydroxide solution. About 2 g of activated charcoal is added to the solid, mixed with water and then heated for about 5 minutes. The alkaline solution is then filtered with suction, and poured into dilute sulphuric acid. The precipitated p-nitrobenzoic acid is filtered at the pump, washed with water and dried.

(Expected yield: 3.5g)

Recrystallisation

About 1g of the crude sample is recrystallised from ethanol or benzene.

(Melting point: 236-238° C)

(07) ANTHRAQUINONE FROM ANTHRACENE

Principle

Anthracene on oxidation using chromic oxide gives anthraquinone.

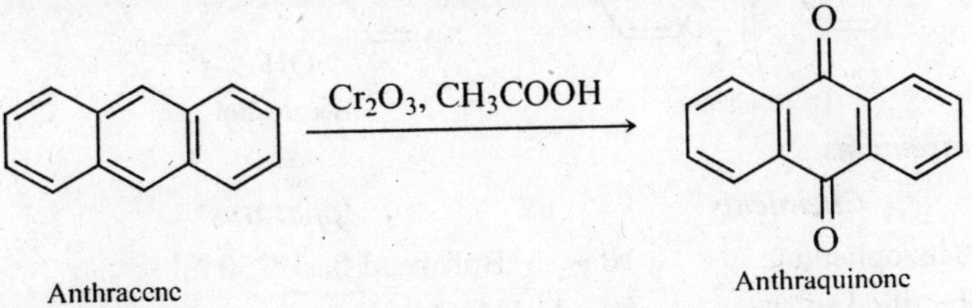

Requirements

Chemicals		***Apparatus***	
Anthracene	2 g	Round bottom flask (250 ml)	1
Glacial acetic acid	40 ml	Reflux condenser	1
Chromium trioxide	4 g	Beaker 250 ml; Funnel	1
		Watch glass, Glass rod	1

Procedure

A mixture of anthracene and 25 ml glacial acetic acid is taken in a round-bottom flask, fitted with a reflux condenser and boiled. The chromic trioxide is mixed with 15 ml of glacial acetic acid and added slowly through the top of the reflux condenser. The solution is continued to boil for an hour steadily. The reaction mixture is then cooled in ice where an yellow precipitate appears. The yellow powder is filtered off at the pump, washed with 5% NaOH solution and then with water and dried.

(Expected Yield: 2g)

Recrystallisation

A small portion of the anthraquinone sample is recrystallised from toluene.

(Melting point: 285° C)

H. REDUCTION

BENZHYDROL FROM BENZOPHENONE

Principle

Reduction of an aromatic ketone with zinc powder and alcoholic sodiur hydroxide, leads to a secondary aromatic alcohol. Hence reduction of benzophenone gives benzhhydrol.

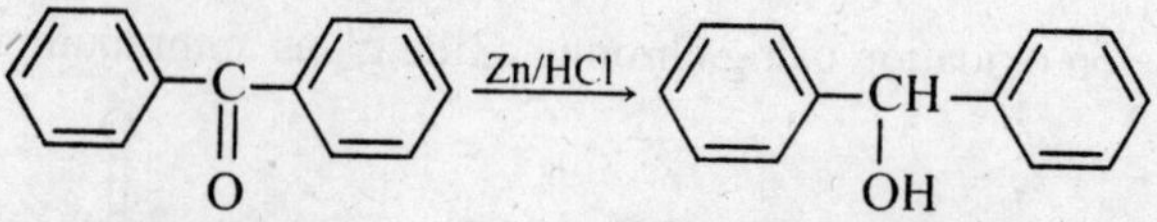

Benzophenone Benzhydrol

Requirements

Chemicals		*Apparatus*	
Benzophenone	10 g	Bolt-head flask 250 ml	1
Rectified spirit	50 ml	Water bath	1
Sodium hydroxide	15 g	Beaker 250 ml	1
Zinc powder	10 g	Funnel, glass rod, watch glass	
Conc. HCl	15 ml		

Procedure

A mixture of benzophenone, rectified spirit, sodium hyroxide and zinc powder is taken in a 250 ml bolt-head flask, fitted with a double surface condenser, and heated in a water bath for about an hour. The heating is continued till the oily layer of benzophenone disappears. The reaction mixture is then allowed to cool in ice where the product separates out. The product is washed with ice cold water acidified with 25ml of concentrated hydrochloric acid. The benzhydrol usually separates as a viscous oil, which crystallizes as a white solid on standing. The white residue is filtered, washed with water and dried.

(Expected yield: 8 g)

Recrystallisation

The white residue of benzhydrol is recrystallised from hot ethyl alcohol.

(Melting point: 68°C)

I. ESTERIFICATION

When a mixture of a carboxylic acid and an alcohol is heated in the presence of a dehydrating agent, esterification takes place.

$$RCOOH + HOR' \xrightarrow{H_2SO_4} RCOOR' + H_2O$$

(01) Dimethyl Oxalate from Oxalic Acid

When oxalic acid is treated with methanol, esterification takes place, resulting in the formation of solid dimethyl oxalate.

$$\underset{\text{Oxalic acid}}{\begin{array}{c}COOH\\ |\\ COOH\end{array}} + 2CH_3OH \longrightarrow \underset{\text{Dimethyl oxalate}}{\begin{array}{c}COOCH_3\\ |\\ COOCH_3\end{array}} + 2H_2O$$

Requirements

Chemicals		***Apparatus***	
Oxalic acid	10 g.	Round-bottom flask (250 ml)	1
Absolute methanol	15 ml	Reflux condenser	1
		Young's still head	1
		Beaker (250 ml)	1
		China dish	1

Procedure

Anhydrous oxalic acid is prepared by heating the hydrated sample to 120°C in an air-oven for about 45 minutes . It is then mixed with 10 ml pure methanol in a round-bottom flask (fitted with a reflux condenser), and heated on a boiling water-bath for about 2 hours. The round-bottom flask is then fitted with Young's still heat (4 bulbs) and the contents fractionally distilled using water condenser first (unreacted methanol distils over). When the temperature reaches about 120°C, the receiver is changed to a China dish and the distillation is continued using air-condenser. The distillate now coming over solidifies on the China dish. It is then dried in air and the yield is noted.

(Expected yield : 3 g.)

Recrystallisation

About 1 g of the crude sample is recrystallised from minimum amount of hot methyl alcohol.

(Correct melting point : 54°C.)

(02) Methyl Salicylate from Salicylic Acid

Principle

Esterification of salicylic acid using methyl alcohol and sulphuric acid gives a pleasant smelling liquid methyl salicylate.

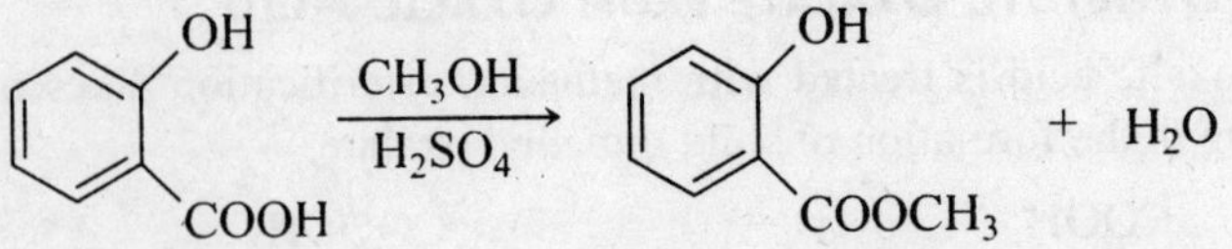

Requirements

Chemicals		Apparatus	
Salicylic acid	8.5 g	Round-bottom flask (100 ml)	1
Methyl alcohol	15 ml	Reflux condenser	1
Concentrated sulphuric acid	2.5 ml	Separating funnel	1
		Water bath	1
		Conical flask, Funnel.	

Procedure

The salicylic acid and methyl alcohol are taken in a 100 ml round-bottom flask and the sulphuric acid added gradually. The mixture is shaken well for about 10 minutes. The flask is fitted with a reflux condenser and the contents are heated on a boiling water-bath for 3 hours. Excess of methyl alcohol is then distilled off on the water-bath. The contents is then transferred to a separating funnel and about 20 ml of water is added to the reaction mixture.

About 30 ml of ether is then added to dissolve the ester formed and the ether layer separated from the aqueous layer. The aqueous layer is then extracted with some more ether. The combined ether extract is then washed repeatedly with dilute sodium carbonate solution till free form acid and finally with water. The ether solution is then dried over anhydrous sodium sulphate. It is then filtered into a dry distilling flask and the ether is removed over a water-bath. The ester left behind is purified by distillation. Methyl salicylate is a colourless liquid with pleasant fragrant smell.

(Expected Yield: 8 g, boiling point: 224°C)

J. DIAZOTISATION

(01) m-NITROPHENOL FROM m-NITROANILINE

Principle

m-Nitroaniline is converted into m-nitrophenol, by first diazotisation followed by heating. On diazotisation, m-nitroaniline is converted into m-nitrophenyldiazonium sulphate below 10°C, which on heating gives m-nitrophenol. This preparation should be done slowly with perfect cooling.

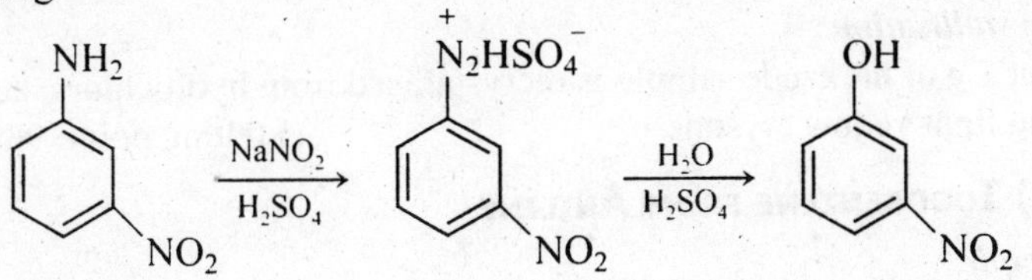

Requirements

Chemicals		*Apparatus*	
m-Nitroaniline	12 g	Beaker 450 ml	1
Sodium nitrite	6 g in 20 ml water	Water bath	1
Conc. Sulphuric acid	20 ml/25 ml water	Round bottom flask	1
Conc. Sulphuric acid	55ml	stirrer, test tube	

Procedure

The concentrated sulphuric acid are cautiously diluted with 25 ml of water in a 450 ml beaker and finely powdered m-nitroaniline is added with stirring to the sulphuric acid. Then crushed ice are added and the mixture stirred until the m-nitroaniline has been converted into the sulphate and a homogeneous paste results. The mass is cooled to 0–5° by placing the beaker in a freezing mixture. A solution of sodium nitrite in water is then added over a period of 10 minutes until a test portion gives a permanent immediate colour with potassium iodide-starch paper. Care is taken to see that the temperature does not rise above 5 to 7° during diazotisation. The stirring is continued for 5 to 10 minutes and the mixture allowed to stand for 5 minutes; some m-nitrophenyldiazonium sulphate may separate. The supernatant liquid is decanted from the solid as far as possible.

While the diazotisation is in progress, 55 ml of concentrated sulphuric acid are added carefully to 50 ml of water in a 500 ml round-bottom flask. The mixture is heated to boiling. The supernatant liquid (diazonium solution) is added from a separating funnel supported over the flask at such a rate that the mixture is maintained at the boiling point.

The residual damp solid is then added in small portions. When all the diazonium salt has been introduced the boiling is continued for a further 5 minutes and the mixture is poured into a 500 ml beaker, cooled in ice water and stirred vigorously. Crystals of m-nitrophenol separate. The crude sample is filtered, washed with cold water and dried.

(Expected Yield: 7g)

Recrystallisation

About 1 g of the crude sample is recrystallised from hydrochloric acid, giving light yellow crystals. (Melting point: 96°C)

(02) IODOBENZENE FROM ANILINE

Principle

Aniline on diazotisation gives benzene diazonium chloride which on treatment with KI gives iodobenzene.

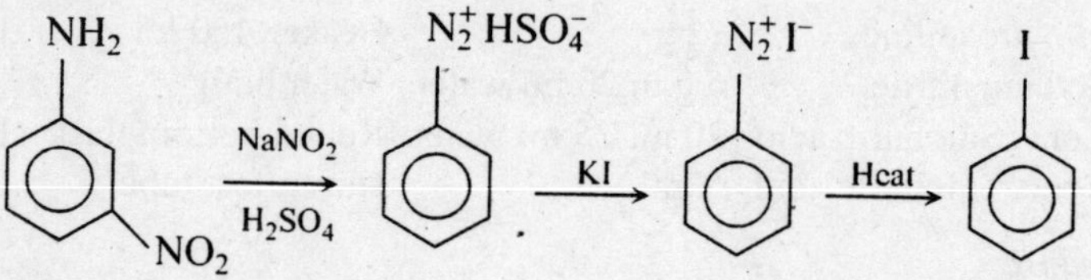

Requirements

Chemicals		Apparatus	
Aniline	10 g	Round bottom flask 250 ml	1
Conc. Sulphuric acid	12 ml	Air condenser	1
Sodium nitrite	8 g/20 ml water	Water bath	1
Potassium iodide	20 g	Test tubes, stirrer	1

Procedure

The aniline is cooled in ice for about 5 minutes and mixed with sodium nitrite solution slowly without allowing the temperature to rise above 10°C and the diazo solution is poured into a litre flask. The potassium iodide dissolved in 50 ml of water is then added and the flask kept on a water-bath heated to about 80°C with an air condenser till the evolution of nitrogen ceases. Iodobenzene separates as a dark heavy oil and is distilled in steam.

The distillate is treated with ether in small portions and the ether extract washed with water at first and then with dilute NaOH solution till all free iodine is removed. The ether extract is dried for about 15 minutes over calcium chloride after which the ether is removed by distillation on a water bath. The iodobenzene is then distilled using an air-condenser and the portion boiling at 188-190°C collected. It is a heavy yellow liquid.

(Expected yield: 18g ; boiling point: 189°C)

(03) p-Chlorobenzoic Acid from p-Toluidine

Principle

p-Toluidine can be converted into p-chlorobenzoic acid by two stages.

In stage-1, p-toluidine is converted into p-chlorotoluene by diazotisation followed by convertion of diazo group into chlorine by Sandmeyer's reaction.

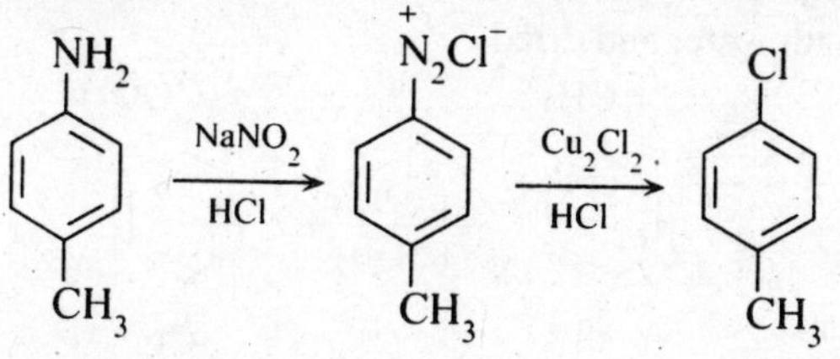

In stage-2, p-chlorotoluene is oxidized into p-chlorobenzoic acid using potassium permanganate.

(a) p-chlorotoluene

Requirements

Chemicals		***Apparatus***
p-Toluidine	15 g	Round bottom flask 250 ml
Sodium nitrite	12 g/20 ml water	Reflux condenser
Con.hydrochloric acid	36 ml/25 ml water	Water bath
Cuprous Chloride	30 g	glass rod

Procedure

The p-toluidine is dissolved in the hydrochloric acid, cooled and diazotised using sodium nitrite solution slowly in a ice-cold solution. The diazo solution is slowly poured into a nearly boiling solution of the cuprous chloride in a round-bottom flask. The mixture is kept hot on a boiling water-bath for half an hour and then steam distilled. The oily distillate is washed with dilute sodium hydroxide, then with water and finally dried over calcium chloride. The liquid is distilled, the portion coming over between 160-164° is collected. It is a colourless liquid.

(Expected yield: 1 g; Boiling point: 162°C)

(b) Oxidation of p-Chlorotoluene to p-Chlorobenzoic Acid

Procedure

About 4 g of the chlorotoluene is oxidized with alkaline potassium permanganate in a 250 ml round-bottom flask fitted with a reflux condenser, the contents of the flask being kept on the brisk boil continuously and by adding the alkaline $KMnO_4$ slowly through the top of the condenser. The oxidation is a slow process and takes about 4-6 hours. When all the oil has practically disappeared and there is a permanent pink colour, the flask is cooled and the acid liberated by passing sulphur dioxide. The chlorobenzoic acid is precipitated, filtered, washed with water and dried.

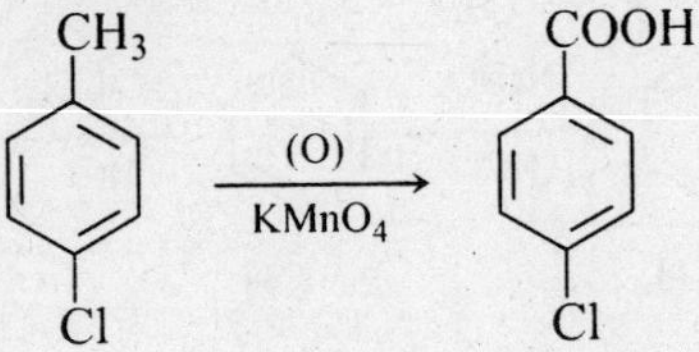

Recrystallisation

About 1 g of the crude sample is recrystallised from dilute alcohol (Melting point: 242°C).

(04) Phenylhydrazine from Aniline

Principle

Aniline is converted into phenyl hydrazine by first diazotisation followed by reduction of benzene diazonium chloride. The reduction can be carried out using tin and HCl or sodium sulphite and sodium carbonate mixture.

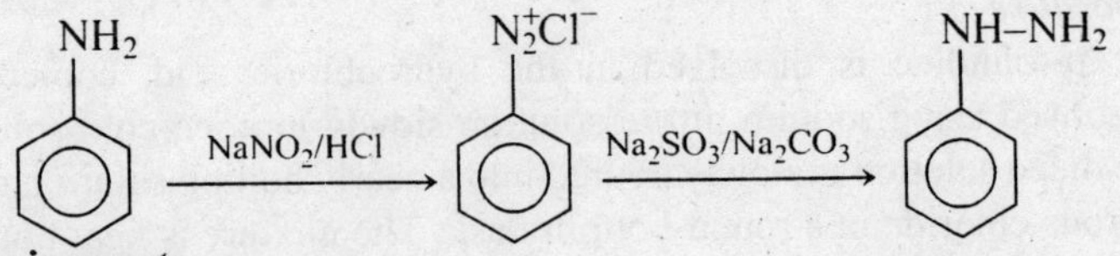

Requirements

Chemicals		*Apparatus*	
Aniline	11 g	Beaker 500 ml	1
Conc. Hydrochloric acid	38 ml/50 ml water	Water bath	1
Sodium nitrite	11 g/20 ml water	Stirrer	1
Sodium sulphite	15g	Glass rod	1
Anhyd. sodium carbonate	12 g		

Procedure

The anhydrous sodium carbonate are dissolved in 50 ml of water. The solution is divided into two equal parts, one part is saturated with sulphur dioxide and the second part then added.

The sodium carbonate is added to the prepared sodium sulphite solution which is kept in a 500 ml beaker well cooled in ice. Aniline is diazotised by adding sodium nitrite in HCl slowly with constant stirring and the diazonium solution is added gradually with constant stirring to the sulphite solution. A deep orange crystalline precipitate is obtained. The mixture is kept stirred for another fifteen minutes and saturated at room temperature with sulphur dioxide to reduce the diazobenzene sulphate.

After saturation, it is warmed to about 70°C by heating on a water bath and filtered. The filtrate is treated with 100 ml of concentrated hydrochloric acid, when the phenylhydrazine hydrochloride separates slowly. The crystallisation is completed by leaving it overnight, after which it is filtered. The hydrochloride is suspended in water and rendered alkaline with caustic soda solution.

The free base liberated as a red oil is extracted twice with 50 ml of benzene. The benzene solution is removed by distillation on water-bath. The residue is distilled under reduced pressure, the portion boiling below 100°C being rejected. Phenylhydrazine boils at 140°C. The portion boiling steadily under the room pressure is collected. It is a light yellow liquid when freshly distilled.

(Expected yield: 11 g; Boiling point: 243°C)

K. ALKYLATION

(01) Anisole from Phenol

Phenol on methylation using dimethyl sulphate in the presence of a base converted into anisole.

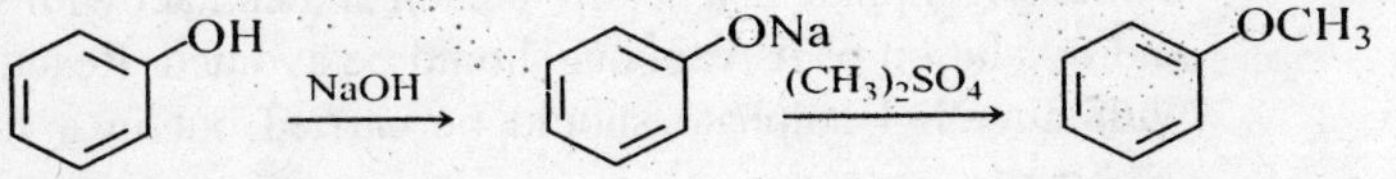

Phenol Anisole

Requirements

Chemicals		Apparatus	
Phenol	10 g	Round-bottom flask (250 ml)	1
Dimethyl sulphate	20 ml	Separating funnel	1
Caustic soda (10%)	50 ml	Thermometer	1
Ether	30 ml	Reflux condenser	1
		Water bath, glass rod.	

Procedure

The phenol is dissolved in the sodium hydroxide solution in a 250 ml round-bottom flask. The dimethyl sulphate is added from a dropping funnel dropwise with continuous stirring. As the methylation proceeds, the temperature rises. The temperature of the mixture is maintained below 50°C by occasional cooling in the tap water. When all the dimethyl sulphate has been added, the reaction mixture is treated with caustic soda solution to make it just alkaline. The stirring is continued for a further half an hour.

The flask is then fitted with a water condenser and heated on a boiling water bath for about an hour. Any excess of dimethyl sulphate will be decomposed by this treatment. The contents of the flask are cooled and transferred into a 250 ml separating funnel and extracted with ether repeatedly atleast three times. All the ether layers are collected. The combined ether extract is dried over anhydrous sodium sulphate. The ether solution is then filtered into a distilling flask.

Purification

The ether is removed over a water bath and the anisole then transferred to a small distilling flask and distilled. It is as colourless oil with a pleasant aromatic odour.

(Expected yield = 10 g, boiling point: 156°C)

Caution

1. Dimethyl sulphate is a deadly poison and contact with skin and inhalation of its vapours should be avoided. Reactions with dimethyl sulphate should be carried out in a fume cupboard.
2. Commercial dimethyl sulphate may contain free acid which should be removed by repeated washing with ice water.

(02) o-NITROANISOLE FROM o-NITROPHENOL

Principle

o-Nitrophenol on methylation using dimethyl sulphate in basic medium gives o-Nitroanisole similar to the convertion of phenol into anisole.

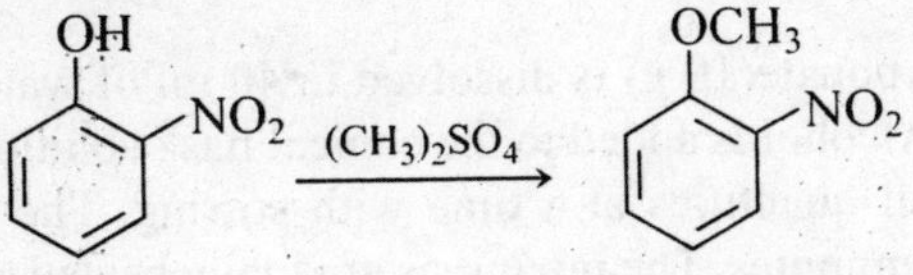

Requirements

Chemicals		*Apparatus*	
o-Nitrophenol	17.5 g	Bolt-heat flask	1
Dimethyl sulphate	24.2 g	Dropping funnel	1
Sodium carbonate (anhydrous)	14.0 g	Reflux condenser	1
Toluene	75 ml	Oil bath, funnel, glass rod	1

Procedure

The o-nitrophenol is mixed with toluene and taken in a one-litre bolt-head flask fitted with stirrer (mercury seal), condenser and dropping funnel. Finely powdered anhydrous sodium carbonate are then added and the mixture heated to 100°C with stirring. Pure dry dimethyl sulphate are added dropwise and after addition is over, the mixture heated at 110-120° for 5-6 hours.

Purification

The mixture is then cooled, and treated with 100 ml of water. To toluene layer is separated and washed with dilute sodium hydroxide solution for removal of unreacted material. The toluene is then distilled off using an oil bath and the residual liquid distilled under reduced pressure.

(Expected yield: 15 g, boiling point: 150°C)

L. MISCELLANEOUS

(01) IODOFORM FROM ETHANOL

Principle

Compounds containing CH3-CO-R or CH3-CHOH-R group like ethanol, acetaldehyde, acetone, ethyl methyl ketone, etc. on treatment with iodine in a basic solution gives an yellow precipitate of iodoform

Requirements

Chemicals		*Apparatus*	
Alcohol (Rectified spirit)	5g	Conical flask 100 ml	1
Potassium carbonate	10 g	Boiling test tube	1
Iodine (powdered)	10 g	Glass rod – 1, beaker, watch glass	

Proceducre

The potassium carbonate(10 g) is dissolved in 40 ml of water in a 100 ml conical flask; Alcohol is added to the conical flask solution and then the iodine in small quantities at a time with stirring. The mixture is shaken well for 10 minutes. The mixture is gradually heated to 80°C and kept there for about five minutes and then cooled. Iodoform separates in small yellow crystals. It is filtered, washed with water, and dried.

(Expected yield: 2.3 g)

Recrystallisation

About 1 g of the sample is recrystallised from alcohol. It forms lemon yellow crystals with a characteristic odour.

(Melting point: 119°C)

(02) Benzamide from Benzoic Acid

Principle

Benzoic acid can be converted into benzamide by first converting into benzoyl chloride followed by the treatment of liquid ammonia.

$$C_6H_5COOH + PCl_5 \rightarrow C_6H_5COCl + POCl_3 + HCl$$

$$C_6H_5COCl + 2NH_4OH \rightarrow C_6H_5CONH_2 + NH_4Cl + 2H_2O$$

Requirements

Chemicals		*Apparatus*	
Benzoic acid	7 g	Distillation flask 100 ml	1
Phosphorous pentachloride	13 g	Beaker 250 ml	1
Liquor ammonia	25 ml	Funnel, glass rod	1

Procedure

Benzoic acid is mixed with the phosphorous pentachloride in a dry 100 ml distilling flask and mixed thoroughly with a glass rod. Dense clouds of hydrogen chloride are given off. When the reaction slackens, the phosphorous oxychloride is distilled off at 107-110°C. The residual liquid which is benzoyl chloride is transferred to a dropping funnel and added drop by drop to 25 ml of liquor ammonia contained in a beaker and cooled in ice water. Heat is evolved during the addition. When the addition is over the liquid is cooled and the precipitated benzamide is filtered, washed with cold water and dried.

(Expected yield: 5 g)

Recrystallisation

About 1 g of crude benzamide is recrystallised from minimum amount of hot water and recrystallised from hot water. It crystallises in colourless plates.

(Melting point: 130°C)

(03) p-Nitroso-N, N-Dimethylaniline

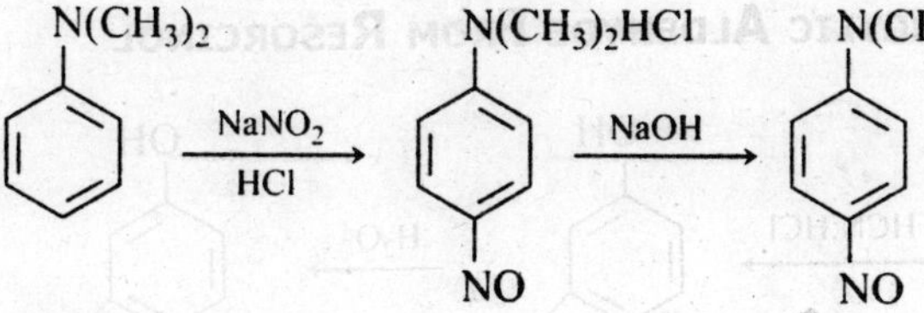

Requirements

Chemicals		***Apparatus***	
N, N-Dimethylaniline	8 ml	Beaker 250 ml	1
Con. Hydrochloric acid	25 ml	Test tubes	3
Sodium nitrite	5 g /10 ml of water	Funnel, Watch glass	

Procedure

The hydrochloric acid diluted with an equal volume of water is taken in a 250 ml beaker and the tertiary base N,N-dimethyl aniline is dissolved in it. The solution is cooled in a freezing mixture of ice and salt (temperature 0-2°C) and the sodium nitrite solution is added in lots of 1 ml at a time with constant stirring. The reaction mixture is tested for the completion of the reaction by taking a drop of the reaction mixture and making contact with starch-iodide paper. The mixture is kept for about half an hour in the ice bath and the yellow crystalline hydrochloride is filtered and washed with ice cold dilute hydrochloric acid.

(Expected Yield: 9 g., Melting point: 177°C)

—

Isolation of the free base

The p-nitroso-N, N-dimethylaniline hydrochloride is made into a thin cream with 10-15 ml of water and transferred to a separating funnel. A 10% solution of sodium carbonate or sodium hydroxide is added with stirring, till the entire mass becomes bright green. The base is extracted with 25 ml of benzene twice and the extract dried over anhydrous potassium carbonate for about 15 minutes. Nearly half the benzene is removed by distillation on a water-bath and the hot solution of the base is transferred to dry porcelain basin for crystallization. Brilliant green leaflets are obtained and they crumble to a fine green crystalline powder after filtration and drying.

(Expected Yield: 8 g., Melting point: 85°C)

(04) Resorcylic Aldehyde From Resorcinol

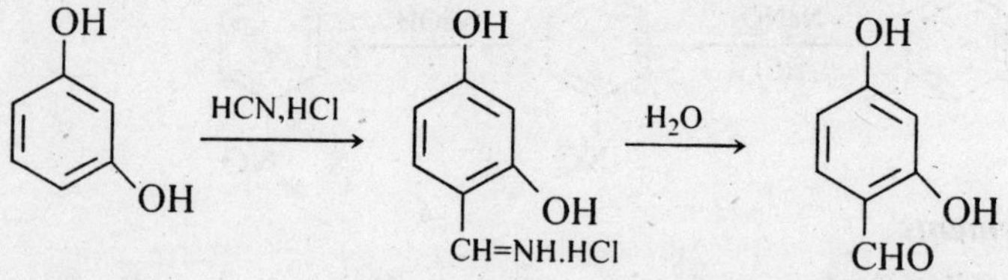

Requirements

Chemicals		Apparatus	
Resorcinol	10 g	Bolt-head flask 500 ml- gas inlet setup	1
Ether (sodium dried)	150 ml	Water bath	1
Zinc cyanide (anhydrous)	14 g	Beaker 250 ml	1
Calcium chloride(anhy)	100g	Funnel, glass rod	
Hydrogen chloride gas			
ice			

Procedure

A bolt-head flask is fitted with a two-holed bung carrying a gas inlet tube and a reflux condenser with a calcium chloride tube attached at the top. Resorcinol, ether and the zinc cyanide are introduced into the flask and the solution is saturated with dry hydrogen chloride while the flask is being cooled in ice. The zinc cyanide gradually disappears and the aldimine hydrochloride separates as a thick oil, which solidifies in about an hour. A slow stream of hydrogen chloride is passed continuously at room temperature for a further period of about 4 hours. At the end of which, the ether is decanted and the solid washed twice with dry ether. The aldimine hydrochloride is decomposed by heating under reflux with 20 ml of water for 5 minutes on a water-bath. The solution is cooled and left overnight. The crystals that separate are filtered, washed with small amounts of water and dried

(Expected yield: 9 g ; Melting point: 135°C)

(05) TETRAHYDROCARBAZOLE FROM CYCLOHEXANONE

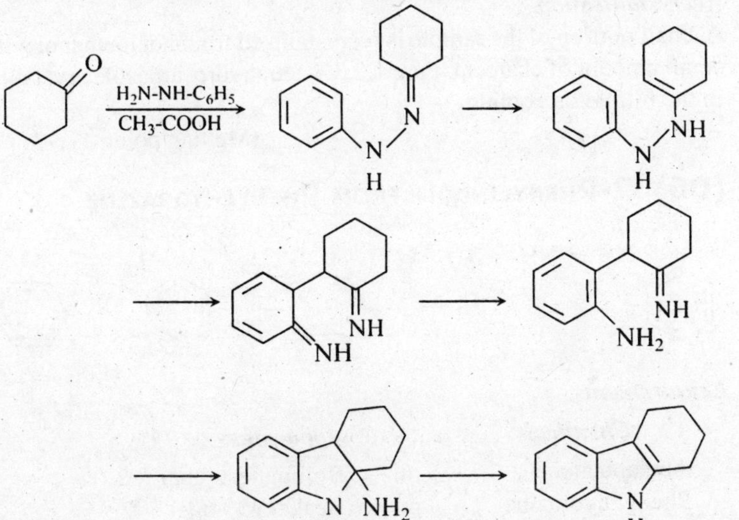

Principle

Cyclohexanone reacts with phenyl hydrazine in glacial acetic acid gives tetrahydrocarbazole. The Fischer indolization reaction is carried out mostly without isolation of the intermediate hydrazone, particularly where the reaction involves an alicyclic ketone.

Requirements

Chemicals		*Apparatus*	
Cyclohexanone	3 g	Round-bottom flask 250 ml	1
Glacial acetic acid	9 ml	Reflux condenser, wire gauze	1
Phenyl hydrazine	3 ml	Beaker 250 ml	1

Procedure

A mixture of cyclohexanone and glacial acetic acid is taken in a round bottom flask fitted with a reflux condenser. The solution is then heated to boil gently on a wire gauze. Phenyl hydrazine is then added slowly through the top of the condenser with constant stirring. The reaction mixture is refluxed for about 30-45 minutes. The disappearance of oily layer indicates the completion of the reaction and the hot mixture is poured into ice taken in a beaker. The content is then stirred vigorously for about 30 minutes. The crude solid is filtered off at the pump, washed with water and dried.

(Expected Yield: 4 g)

Recrystallisation

A small portion of the sample is recrystallised from hot methanol with a small amount of charcoal. Pure 1,2,3,4-tetrahydrocarbazole crystallizes in the filtrate on cooling.

(Melting point: 116-117°C)

(06) 2-PHENYLINDOLE FROM PHENYLHYDRAZINE

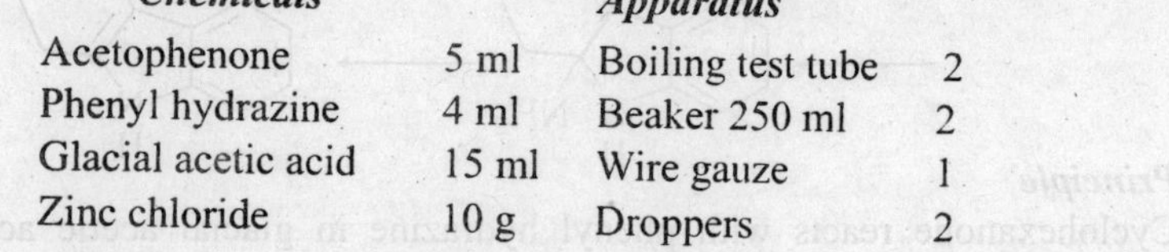

Requirements

Chemicals		***Apparatus***	
Acetophenone	5 ml	Boiling test tube	2
Phenyl hydrazine	4 ml	Beaker 250 ml	2
Glacial acetic acid	15 ml	Wire gauze	1
Zinc chloride	10 g	Droppers	2

Procedure

A mixture of acetophenone, phenyl hydrazine and 2 drops of glacial acetic acid is taken in a boiling tube and shaken well for 5 minutes where crystals of phenylhydrazone separates out.

A powdered anhydrous zinc chloride is added to the mixture in a beaker and heated to 130-140°C. A ring closure reaction taken place and during this process, the heat should not be allowed to rise beyond 180°C. The reaction mixture is kept hot for about 15 minutes. The mixture is then cooled slowly and about 15 ml of glacial acetic acid is added and heated for 10 minutes. The hot mixture is then poured into a beaker and allowed to crystallize. The crystals are filtered at the pump, washed with water and dried.

(Expected yield: 5g)

Recrystallisation

About 1 g of the sample is recrystallised from aqueous ethanol.

(Melting point: 189°C)

(07) DINITROPHENOL

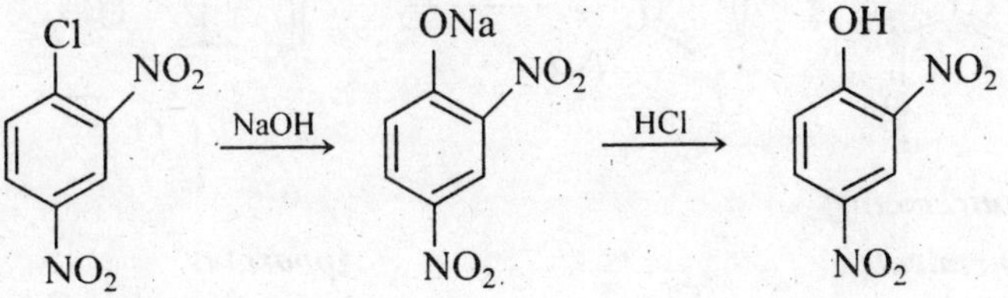

Requirements

Chemicals		Apparatus	
2,4-dinitrochlorobenzene	10 g	Round-bottom flask	1
Sodium carbonate (anhydrous)	15 g	Reflux condenser	1
Or Sodium hydroxide	10 g	Funnel	1
Dilute HCl	30 ml	glass rod	1

Procedure

The sodium carbonate or hydroxide is dissolved in 100 ml of water contained in a round bottom flask. The 2,4-dinitrochlorobenzene is added and the mixture is heated using a water-cooled condenser till the oily layer disappears, and the liquid becomes clear. It may take about 2-3 hours. The clear liquid is filtered hot using a Buchner funnel and the filtrate acidified with hydrochloric acid. On cooling, the dinitrophenol separates. It is filtered, washed with cold water and dried.

(Expected yield: 8 g)

Recrystallisation

The crude crystals of dinitrophenol is recrystallised from minimum amount of hot water.

(Melting point: 112°C)

(08) FLUORESCEIN

Principle

Fluorescein is prepared by heating phthalic anhydride or phthalic acid and resorcinol with con. sulphuric acid or anhydrous zinc chloride. Generally anhydrous zinc chloride is preferred to prepare fluorescein in the solid state.

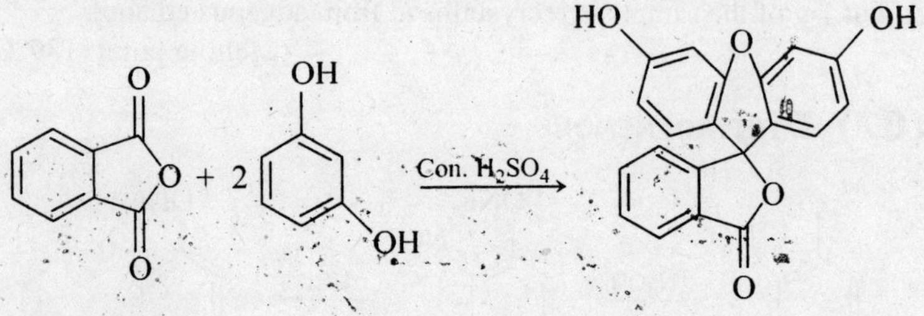

Requirements

Chemicals

Resorcinol	5 g
Phthalic anhydride	4 g
Concentrated sulphuric acid	3 g
Anhydrous $ZnCl_2$	2 g
Con. HCl	5 ml

Apparatus

Round bottom flask (250 ml)	1
Beaker 400 ml	1
Filtering flask, funnel	1
Glass rod	1

Procedure

An intimate mixture of phthalic anhydride and resorcinol is heated to 180°C in a 250 ml round-bottom flask placed in an oil-bath. Powdered anhydrous zinc chloride are added in small portions during 30 minutes, stirring with the glass rod. At the end, the mass becomes so viscous that stirring is not possible. The oil bath is allowed to cool to 90^0 and a mixture of 100 ml water and 5 ml concentrated hydrochloric acid is added to the reaction mixture. The temperature of the oil bath is raised until the water boils. The mixture is stirred, till the reaction mass disintegrates and the zinc salts have dissolved. The mass is filtered, the residue ground with water in a mortar, filtered and dried at 100°.

(Expected Yield: 3 g)

(09) EOSIN

Principle

When fluorescein is mixed with bromine in alcohol, tetrabromofluorescein is formed even in cold condition, called eosin.

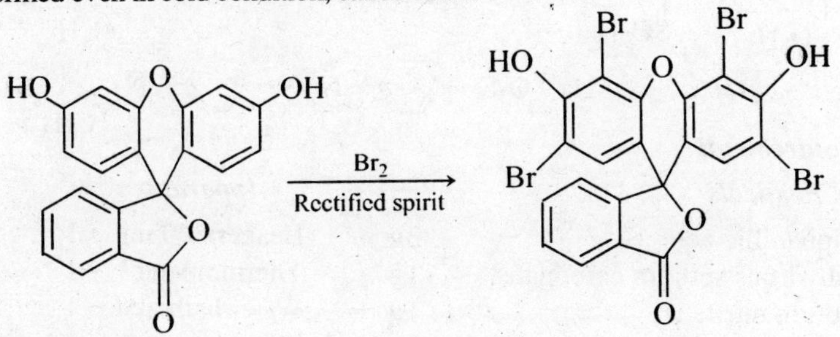

Requirements

Chemicals		*Apparatus*
Fluorescein	5 g	Round bottom flask 250 ml
Bromine	10 g	beaker 400 ml, Separating funnel
Rectified spirit	20 ml	Filtering flask, funnel
		Glass rod

Procedure

A mixture of powdered fluorescein and rectified spirit are kept in a round bottom flask and bromine is added dropwise from a separating funnel. When half the bromine has been added, a clear solution results and crystalline material separates out on further addition of bromine. The reaction mixture is allowed to stand for 1 hour and the eosin is filtered and washed with alcohol and dried at 100°C.

(Expected yield: 5 g.)

M. DYES

(01) METHYL ORANGE FROM SULPHANILIC ACID

Principle

Methyl orange is prepared by coupling diazotised sulphanilic acid with N,N-dimethyl aniline in the presence of excess hydrochloric acid.

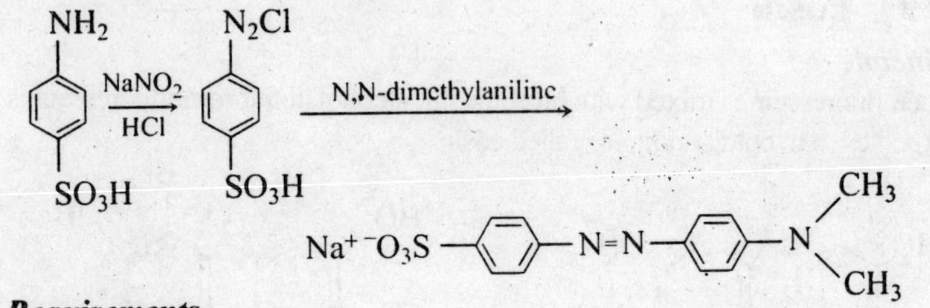

Requirements

Chemicals		*Apparatus*	
Sulphanillic acid	3g	Beaker 500 ml	1
Anhydrous sodium carbonate	1g	Thermometer	1
Sodium nitrite	1g	Water bath, ice	1
Dimethyl aniline	2ml	Stirrer	1
Sodium chloride	8g	Glass rod	1
Con. HCl	3 ml	Funnel	1
NaOH (20% solution)	15 ml		
NaCl	8 g		
Rectified spirit	20 ml		

Procedure

The sodium carbonate is dissolved in 50 ml of water and sulphanilic acid is added and dissolved by boiling the solution. The mixture is cooled in ice and sodium nitrite in water is added to the contents. To the ice cold solution conc. hydrochloric acid is added drop by drop with constant stirring and the temperature being maintained between 0-5°C. To the diazotised solution added an ice cold solution of dimethyl aniline in 10 ml dilute hydrochloric acid with continuous stirring. The solution is made alkaline by adding 15 ml of 20% sodium hydroxide solution and the dye is salted out by adding sodium chloride. The mixture is heated to boiling, cooled in an ice bath. The orange coloured dye is filtered, washed with little ethanol and dried.

(Expected yield: 6 g)

(02) Methyl Red

Principle

Anthranilic acid on diazotisation followed by coupling with N,N-dimethylaniline gives methyl red, a dye.

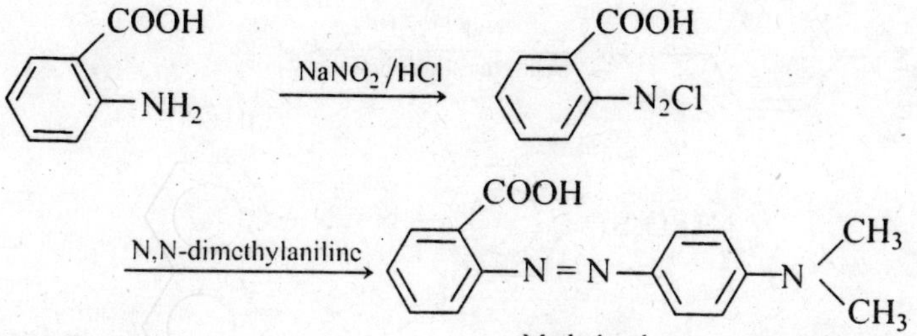

Methyl red

Requirements

Chemicals		*Apparatus*	
Anthranilic acid	5 g	beaker 250, 400 ml	2
Con.hydrochloric acid	2 ml	Separating funnel	1
Sodium nitrite	2 g	Filtering flask, funnel	1
N, N-Dimethylaniline	2 g	Glass rod	1
Sodium acetate	10 g	Water bath	1

Procedure

Into a 250 ml beaker, 50 ml of water is taken, the hydrochloric acid is added first and then the anthranilic acid. The mixture is stirred till solution is complete, cooled in ice and then diazotised by adding in small portions the solid nitrite. The solution is kept in ice for half an hour. Dimethylaniline is dissolved in water containing 5 ml of con. hydrochloric acid in another beaker. The diazonium salt is poured into this solution and the sodium acetate added immediately. The mixture is warmed to 40°C on a water bath, when the red dye separates quickly. After an interval of three hours by which time the reaction is complete, the dye is filtered, washed with water freely as it is insoluble, and air dried. It is difficult to crystallise.

(Expected yield: 3 g.)

(03) Orange II

Principle

Diazotisation of sulphanilic acid following by coupling with α-naphthol in alkaline medium gives organge II, a dye.

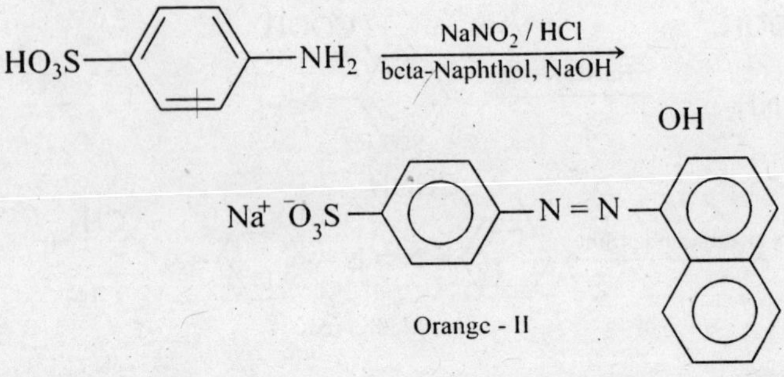

Orange - II

Requirements

Chemicals		*Apparatus*	
Sulphanilic acid	5 g		
Con.hydrochloric acid	2 ml	beaker 250, 400 ml	2
		Separating funnel	1
Sodium nitrite	2 g	Filtering flask, funnel	1
β-Naphthol	2 g	Glass rod	1
Sodium chloride	10 g	Water bath	1

Procedure

Into a 250 ml beaker, 50 ml of water is taken, the hydrochloric acid is added first and then the sulphanilic acid. The mixture is stirred till solution is complete, cooled in ice and then diazotised by adding in small portions the solid nitrite. The solution is kept in ice for half an hour. β-Napthol is dissolved in water containing 20 ml of 10% NaOH solution in another beaker. The diazonium salt is poured into this solution and the sodium chloride added immediately. The dye which crystallises on cooling to room temperature, is filtered and washed with a little saturated salt solution and dried at 80°. The pure dye which crystallises out is filtered, washed with alcohol and dried in air.

(Expected yield: 5 g.)

(04) p-Nitrophenylazo-β-Naphthol

Principle

p-Nitroaniline on diazotisation followed by coupling with β-naphthol in alkaline medium gives the azo dye, p-nitrophenylazo-β-naphthol.

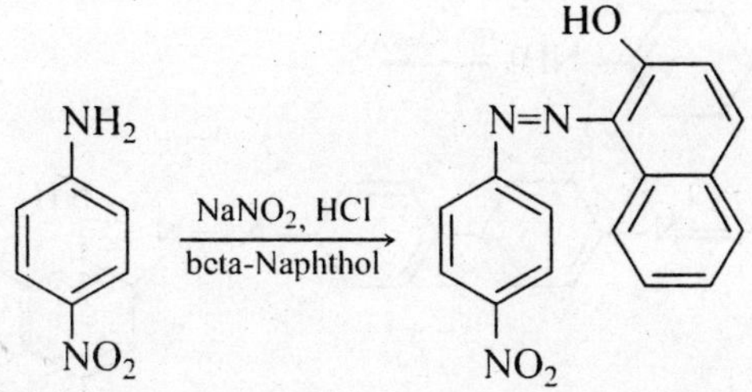

Requirements

Chemicals		*Apparatus*	
p-Nitraniline	5 g	Dropping funnel	1
Con.sulphuric acid	15 ml	beaker 250, 400 ml,	2
Sodium nitrite	5 g	Separating funnel	1
β-Naphthol	5 g	Filtering flask, funnel	1
Sodium acetate	5 g	Glass rod	1
Sodium hydroxide(10%)	30 ml	Water bath	1

Procedure

The p-nitroaniline is mixed with the diluted acid in a 250 ml beaker and cooled in a freezing mixture of ice and salt. The nitrite solution is added slowly and carefully from a dropping funnel. At first the p-nitraniline dissolves, and a pasty mass of the diazo compound separates. The sodium acetate solution is then added, and the solution of β-naphthol in alkali is diluted to 50 ml in a 400 ml beaker. The diazo solution is then mixed with stirring. Coupling takes place and a scarlet precipitate of the dye is obtained. After keeping for about ten minutes, the dye is filtered at the pump, washed with hot water and dried.

(Expected yield: 6 g.)

(05) Congo Red

Principle

Benzidine on diazotisation followed by coupling with sodium naphthionate in alkaline solution gives the dye congo red.

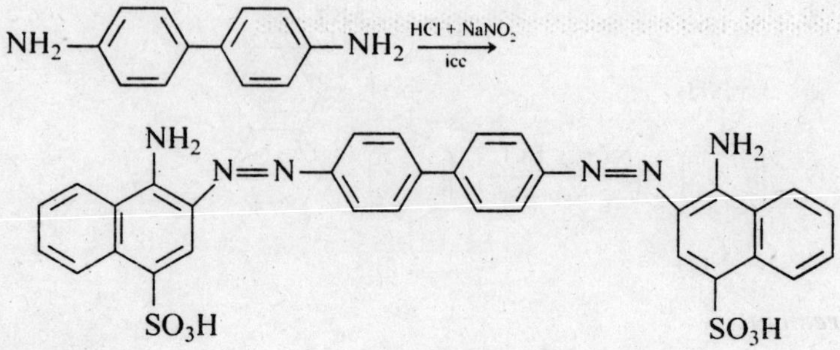

Requirements

Chemicals		*Apparatus*	
Benzidine	5 g		
Con. HCl	15 ml	Beaker 250, 400 ml	2
Sodium nitrite	5 g	Separating funnel	1
Sodium naphthionate	10 g	Filtering flask, funnel	1
Sodium acetate	5 g	Glass rod	1
Sodium carbonate	10 g	Water bath	1
Sodium hydroxide(10%)	30 ml		

Procedure

Benzidine is dissolved in a hot mixture of concentrated hydrochloric acid and 50 ml of water. The solution is cooled in an ice bath and treated with a solution of sodium nitrite in water with good stirring. This is called tetrazo solution. After few minutes, the tetrazo solution is added to a solution of sodium naphthionate and crystallised sodium acetate in 250 ml of water. The mixture is stirred well and let stand for one hour. The coupling is complete, when a test portion, on warming with hydrochloric acid, does not evolve nitrogen. The blue-black precipitate of the dye acid is dissolved by the gradual addition of sodium carbonate followed by stirring and warming to 80°C. The hot solution is filtered, the filtrate just saturated with sodium chloride, allowed to cool to room temperature and then cooled in ice. The congo red is filtered and washed with saturated solution of sodium chloride and dried in air.

(Expected yield: 10 g).

(06) MALACHITE GREEN

Principle

Malachite green is a condensation product obtained from benzaldehyde and N,N-dimethylaniline in acid medium. The product obtained is converted into a dye by oxidation.

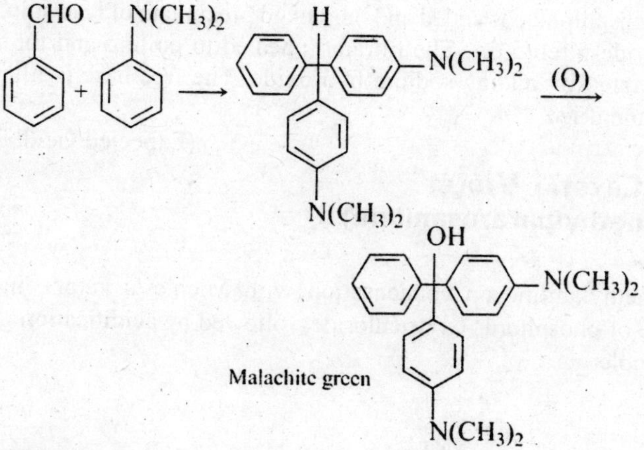

Requirements

Chemicals		*Apparatus*	
N,N-Dimethylaniline	5 g	Roundbottom flask	1
Con. HCl	15 ml	beaker 250, 400 ml	2
Sodium nitrite	5 g	Separating funnel	1
Benzaldehyde	5 g	Filtering flask, funnel	1
Sodium acetate	5 g	Glass rod	1
Sodium hydroxide(10%)	30 ml	Water bath	1
Sodium carbonate	10 g		
Lead peroxide	10 g		

Procedure

Freshly distilled dimethylaniline, benzaldehyde and concentrated hydrochloric acid are introduced into a round bottom flask fitted with a water-cooled condenser and heated for about 2 hours. The reaction product is rendered alkaline with caustic soda and distilled in steam to remove the unreacted benzaldehyde and dimethylaniline. The residue is poured into 200 ml of water and the leuco base is obtained in the form of granules. It is filtered and washed free from alkali and dried at 100°C. The leuco base is converted into the dye by oxidation. The base is dissolved by gentle warming in dilute hydrochloric acid and 4 g of glacial acetic acid. The liquid is diluted to 400 ml and cooled by adding crushed ice. Pure lead peroxide are added as a thin paste and shaken frequently for almost 10 minutes after which a saturated solution of 10 g of sodium sulphate is added and the mixed precipitate of lead sulphate and chloride filtered off. The filtrate is heated to boiling and the free base liberated by adding sodium hydroxide. The free base is filtered, washed and dried.

(Expected yield, 8 g.)

(07) CRYSTAL VIOLET (Hexamethylpararosaniline)

Principle

N,N-dimethylaniline on condensation with Michler's ketone in the presence of phosphorus oxytrichloride, followed by acidification gives crystal violet.

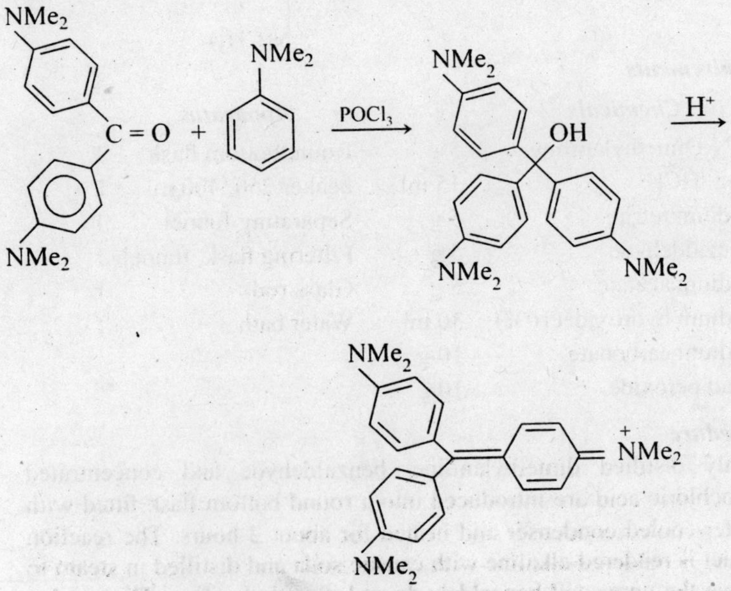

Crystal violet

Requirements

Chemicals		*Apparatus*	
N,N-Dimethylaniline	10 g	Round bottom flask	1
Con. HCl	5 ml	Beaker 250, 400 ml	2
Michler's ketone	5 g	Separating funnel	1
Phosphorus oxytrichloride	5 g	Filtering flask, funnel	1
Sodium chloride	20 g	Glass rod	1
Sodium hydroxide(10%)	30 ml	Water bath	1

Procedure

A mixture of dimethylaniline, Michler's ketone and phosphorus oxytrichloride are introduced into a round bottom flask and heated for about four hours by immersing the flask in a vigorously boiling water bath. After about half an hour the mixture solidifies to a cake and the heating is continued. About 50 ml of water are poured into the flask and the blue solution rendered alkaline with a solution of caustic soda. The flask is cooled, the reddish solid separated by filtration and boiled with a mixture of 50 ml of water to which only 2 ml of concentrated hydrochloric acid have been added. The blue solution is allowed to settle and filtered through a folded filter paper. Any residue is treated similarly, using 50 ml of the same acid solution till the whole of it goes into solution. The combined filtrate is cooled and treated while stirring well, with finely powdered common salt until the dye is precipitated. The dye is filtered and recrystallised from boiling water. Crystal violet separates from the solution in fine crystals have a bronze-like lustre, which are dried in air.

(Expected yield: 4 g.)

N. Preparations Involving Name Reactions

(01) Benzoin Condensation Reaction (Benzoin from Benzaldehyde)

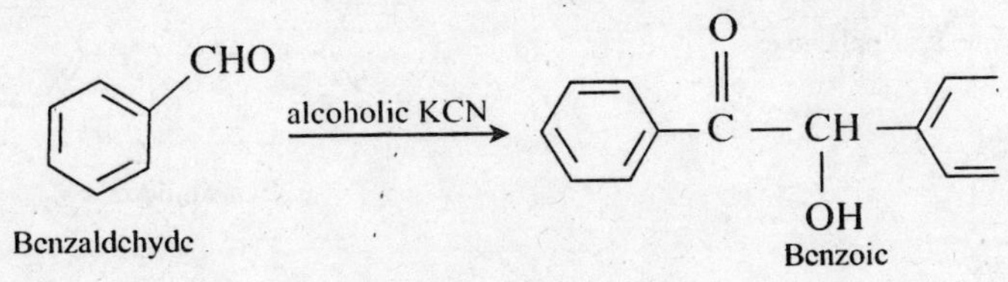

Principle

When benzaldehyde is heated with alcoholic KCN solution, a condensation takes place which gives benzoin, a aromatic hydroxy ketone.

Requirements

Chemicals		Apparatus
Benzaldehyde	20 g	Round-bottom flask 250 ml
Potassium cyanide	4g/15 ml water	Reflux condenser
Rectified spirit	40 ml	Water bath, Beaker, Glass rod, Funnel

Procedure

The mixture of benzaldehyde, potassium cyanide solution and alcohol is taken in a round-bottom flask, fitted with a Leibig condenser and refluxed for about an hour. On cooling, benzoin separates as a crystalline mass. The white solid that separates out is filtered, washed with a little alcohol and dried.

Recrystallisation

About 1 g of crude benzoin are recrystallised from rectified spirit. It crystallises in colourless prisms.

(Expected Yield: 15 g; Melting point: 133°C)

Caution: Potassium cyanide is highly poisonous. As it is a deadly poison, its contact with skin or face should be avoided. A pair of gloves should be used and the reaction may be carrect out in fume chamber.

(02) Beckmann Rearrangement (Benzanilide from Benzophenone)

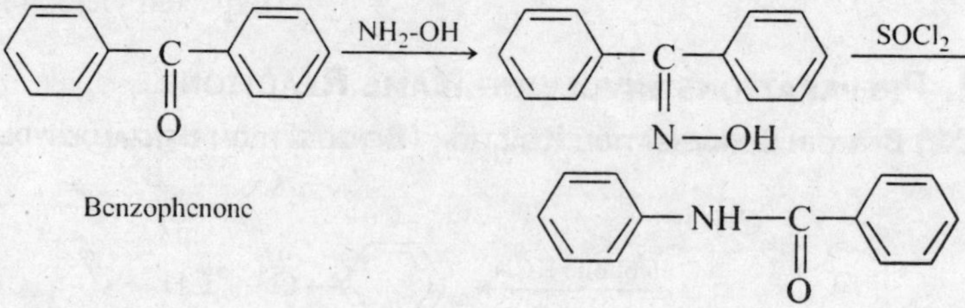

Benzophenone

Benzanilide

Principle

When benzaldehyde is condensed with hydroxylamine, one water molecule is removed and gives benzophenoxime or benzophenone oxime. This undergoes Beckmann rearrangement followed by hydrolysis gives benzanilide.

Requirements

Chemicals		*Apparatus*	
Benzophenone	5 g	Round-bottom flask 250 ml	1
Hydroxylamine HCl	3 g	Reflux condenser	1
Rectified spirit	50 ml	Water bath	1
Sodium hydroxide	8 g	Beaker 250 ml, Glass rod, Funnel	
Con. HCl	15 ml		

(a) Benzophenone Oxime

A mixture of benzophenone, hydroxylamine hydrochloride, about 10 ml water and rectified spirit is placed in a round-bottom flask. Sodium hydroxide pellets are then added in portions with shaking to the mixture in the flask. When the reaction becomes too vigorous the flask is cooled with tap water. When all the sodium hydroxide has been added, the flask is fitted with a reflux condenser and heated on a boiling water bath for about 15 minutes. The flask is then cooled, the contents poured into 100 ml of water containing conc. HCl. The precipitate is filtered at the pump, washed with water and dried in a vaccum desiccator over sulphuric acid.

(Expected Yield: 5 g; Melting point: 142°C)

(b) Benzanilide

Requirements

Chemicals		*Apparatus*	
Benzophenone oxime	2 g	Round-bottom flask 250 ml	1
Anhydrous ether	20 ml	Test tube	1
Thionyl chloride	3 ml	Glass rod, Funnel	1
Rectified spirit	10 ml		

A solution of benzophenone oxime in anhydrous ether is treated with pure thionyl chloride in a round bottom flask. The solvent is then distilled off and the residue treated with a test tube full of water. The mixture is boiled for a few minutes, any lumps formed being broken up with a glass rod. The supernatant liquid is decanted off and the residue is filtered, washed with cold water and dried.

Recrystallisation

About 1 g of the crude sample is recrystallised from boiling alcohol.

(Expected Yield: 4 g; Melting point: 163°C)

(03) Knoevenagel Reaction

(p–Methoxycinnamic Acid from Anisaldehyde)

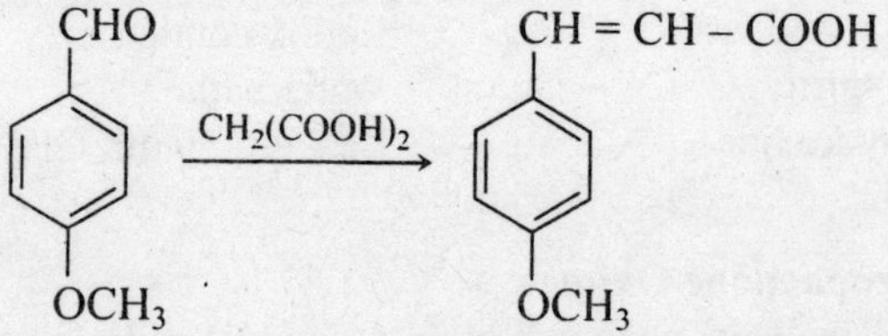

Principle

Aromatic aldehydes like benzaldehyde, anisaldehyde, etc. react with malonic acid in basic medium give substituted cinnamic acid. This reaction is known as Knoevenagel reaction. This reaction takes place in the presence of pyridine also.

Requirements

Chemicals		*Apparatus*	
Anisaldehyde (freshly distilled)	13 g	Conical flask 250 ml with a cork	1
Malonic acid	12 g	Water bath	1
Ammonia (8%) in ethyl alcohol	21 g	Beaker 250 ml	1
Absolute alcohol	30 ml	Glass rod	1
Dilute sulphuric acid	20 ml	Funnel, Watch glass.	1

Procedure

The anisaldehyde and malonic acid are dissolved in the minimum amount of absolute alcohol in a 250 ml conical flask and the alcoholic ammonia added to the solution. The flask is corked and left to stand for 10–15 minutes. The alcohol is then distilled off on a water bath and the oily residue in the flask heated on a vigorously boiling water bath till the evolution of carbon dioxide ceases and the mixture becomes solid. This may take about 1½ hours to 2 hours. The residue is then dissolved in minimum amount of sodium carbonate solution and the resulting liquid boiled and filtered. The warm filtrate is poured into an excess of 20 ml sulphuric acid containing crushed ice. The precipitated acid is filtered, washed with water and dried.

(Expected Yield: 8g)

Recrystallisation

About 1 g of the sample is recrystallised from boiling water or rectified spirit.

(Melting point: 172°C)

(04) SANDMEYER'S REACTION
(p-CHLOROTOLUENE FROM p-TOLUIDINE)

Principle

p-Toluidine on diazotisation gives diazonium salt which on treatment with cuprous chloride and HCl gives p-chlorotoluene. Conversion of diazonium salt into chloro compound using cuprous chloride is an example of Sandmeyer's reaction.

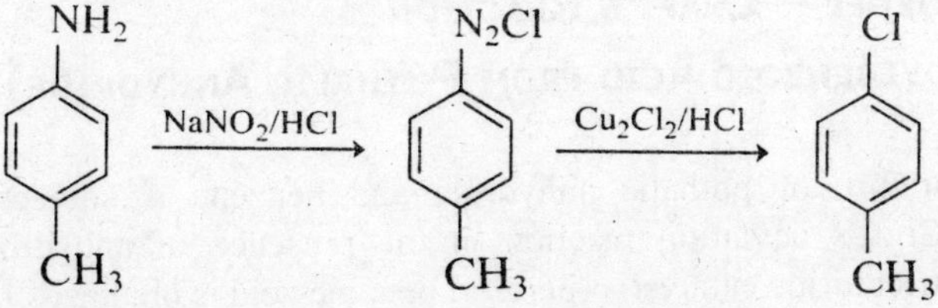

Requirements

Chemicals		***Apparatus***
p-Toluidine	4 g	Conical flask 250 ml
Conc. HCl	10 ml	Water bath
Sodium nitrite	3 g	Dropping funnel
Cuprous chloride	5 g	

Procedure

A mixture of p-toluidine, about 10 ml water and concentrated hydrochloric acid are taken in a Conical flask and the solution is cooled in ice perfectly. Sodium nitrite is dissolved in water and added to the conical flask solution slowly with constant stirring without allowing the temperature to rise. The completion of the reaction and the presence of sodium nitrite in excess may be ascertained by testing with starch-iodide paper which gives blue colouration with nitrite salt. The cuprous chloride is dissolved in water and this solution is slowly added to the conical flask solution with vigorous shaking till the mixture becomes highly viscous. This reaction mixture is allowed to stand at room temperature for about 30 minutes where oily layer of p-chlorotoluene is formed. The mixture is warmed gently on a water bath for about 30 minutes with shaking. The distillate is steam distilled to get the product. The mixture is then taken in a separating funnel and 10 ml of 10% NaOH solution are added, shaken well and the top organic layer is removed. The top layer is further washed with NaOH solution and water thoroughly and then with dilute sulphuric acid followed by water.

(Expected yield: 4g)

Purification

The product is dried over anhydrous magnesium sulphate and redistilled.

(Boiling point: 160°C)

(05) Friedel – Craft's Reaction
(o-Benzoylbenzoic Acid from Phthalic Anhydride)

Principle

When a mixture of phthalic anhydride and benzene is subjected to Friedel-Craft's acylation reaction in the presence of anhydrous aluminium trichloride catalyst, o-benzoyl benzoic acid is obtained.

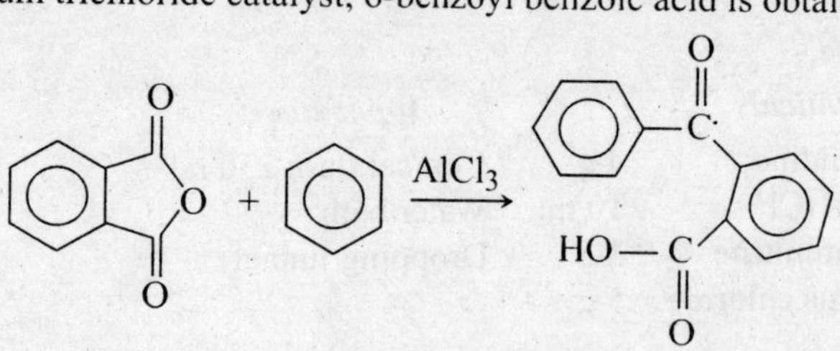

Requirements

Chemicals		*Apparatus*	
Phthalic anhydride	10g	Bolt-head flask 250 ml	1
Anhy. Aluminium trichloride	15 g	Beaker 250ml	1
Pure Benzene	15 g	Funnel, Watch glass, Glass rod.	

Procedure

Into a dry bolt-head flask with a double surface condenser, a mixture of phthalic anhydride, sodium dried benzene and powered anhydrous aluminium chloride is taken and mixed thoroughly for about 20 minutes. The mixture is heated on a boiling water bath for about an hour till the evolution of hydrogen chloride gas is completely ceased. The flask is cooled in ice with occasional shaking. The excess of benzene is removed by steam distillation . The residue obtained on cooling is filtered off at the pump, washed with water and dired.

(Expected yield: 8 g; Melting point: 128°C)

(06) CLEMMENSEN REDUCTION (ETHYL RESORCINOL)

Principle

Carbonyl compounds on reduction using zinc amalgam and con. HCl give alkane in which the –CO– group is converted into $-CH_2-$ group. This reduction is called Clemmensen reduction

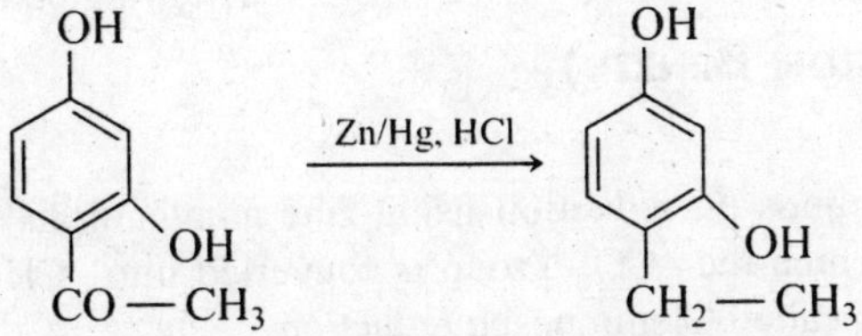

Requirements

Chemicals		*Apparatus*	
Resacetophenone	10 g	Round bottom flask 250 ml	1
Zinc amalgam from zinc	40 g	Beaker 250 ml	1
Concentrated hydrochloric acid	30 ml	Test tubes, Watch glass	
Mercuric chloride	4 g		

Procedure

Zinc amalgam is used as a reducing agent in Clemmensen reduction. This can be prepared as follows: About 40 g of zinc are placed in a round-bottom flask and washed thrice with warm 1:1 hydrochloric acid. 40 ml of water, 2 to 4 g mercuric chloride and 2 ml of concentrated hydrochloric acid are added to the washed zinc and vigorously stirred for about ten minutes. The aqueous solution is decanted off, the amalgamated zinc is washed once with distilled water and covered with 30 ml water and an equal volume of concentrated hydrochloric acid.

10 g of resacetophenone are added to the amalgam contained in the flask and the mixture is refluxed for about 4 hours, 5 ml of concentrated hydrochloric acid being added every hour. At the end of each hour a drop of the liquid is taken in 1 ml ethyl alcohol and tested with aqueous ferric chloride. The completion of reaction is indicated by no colour formation.

When the reaction is over, the contents of the flask are cooled, and the supernatant liquid carefully decanted from the unreacted amalgam. The solution is then saturated with solid sodium chloride, cooled and extracted repeatedly with ether. The combined ether extracts are dried over an hydrous sodium sulphate. On removal of the ether, a light yellow solid is obtained.

(Expected yield: 10g)

Recrystallisation

About 1 g of the crude sample is crystallised from benzene or chloroform. Ethyl resorcinol crystallises as white glistening prisms.

(Melting point : 97°C)

(DIBENZYL FROM BENZIL)

Principle

Carbonyl compounds on reduction using zinc amalgam and con.HCl give alkane in which the –CO– group is converted into $-CH_2-$ group. This reduction is called Clemmensen reduction

$$C_6H_5-CO-CO-C_6H_5 \xrightarrow{\text{Zinc amalgam/HCl}} C_6H_5-CH_2-CH_2-C_6H_5 + 2H_2O$$

Requirements

Chemicals		***Apparatus***	
Benzil	7 g	Round bottom flask 250 ml-1	
Amalgamated Zinc	30 g	Air condenser	1
Hydrochloric acid (1:1)	100 ml	Beaker 250 ml	1
Conc hydrochloric acid	20 ml	Glass rod, Funnel, Watch glass	

Procedure

The benzil, hydrochloric acid and the amalgamated zinc are introduced into a round-bottom flask fitted by a ground joint to a long air-condenser and the mixture kept boiling gently for about 4 hours. 5 ml of concentrated hydrochloric acid are added at the end of each hour. The clear liquid is decanted into a beaker and the reaction product which solidifies is separated by filtration and washed with water repeatedly and dried.

(Expected yield: 5 g)

Recrystallisation

About 1 g of the crude sample is recrystallised from alcohol.

(Melting point: 52°C).

(07) CLAISEN REARRANGEMENT (2-ALLYLPHENOL)

Principle

o-Allyl phenyl ether on heating rearranges into o-allyl phenol. This rearrangement is called Claisen Rearrangement. Allyl phenyl ether is prepared by treating allyl bromide with phenol. This reaction takes place in two steps. In stage-1, phenol is converted into allyl phenyl ether and in Stage-2, allyl phenyl ether is converted into o-allyl phenol

Stage – 1: Preparation of allyl phenyl ether:

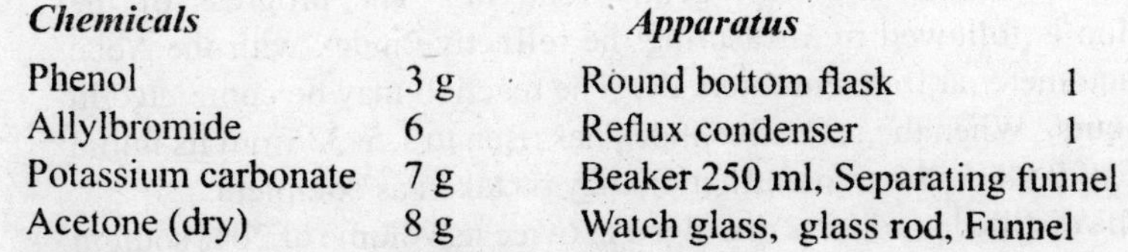

Requirements

Chemicals		*Apparatus*	
Phenol	3 g	Round bottom flask	1
Allylbromide	6	Reflux condenser	1
Potassium carbonate	7 g	Beaker 250 ml, Separating funnel	
Acetone (dry)	8 g	Watch glass, glass rod, Funnel	
Ether	100 ml	Water bath	1
NaOH (2N)	25 ml		

Procedure

The phenol, allyl bromide, potassium carbonate and acetone are introduced into a round-bottom flask fitted with a double surface water-cooled condenser and heated on a boiling water bath for about 8 hours. The contents of the flask are then transferred to a separating funnel and extracted repeatedly with ether, after addition of 50 ml water. The ether extract is washed twice with 2N sodium hydroxide to remove any unreacted phenol and then with water to free it from alkali and dried over anhydrous potassium carbonate. After removal of ether, the residue, which is allylphenyl ether, is purified by distillation under reduced pressure.

(Expected yield: 5 g, Boiling point. 85°C)

Stage – 2: Preparation of o-allyl phenol:

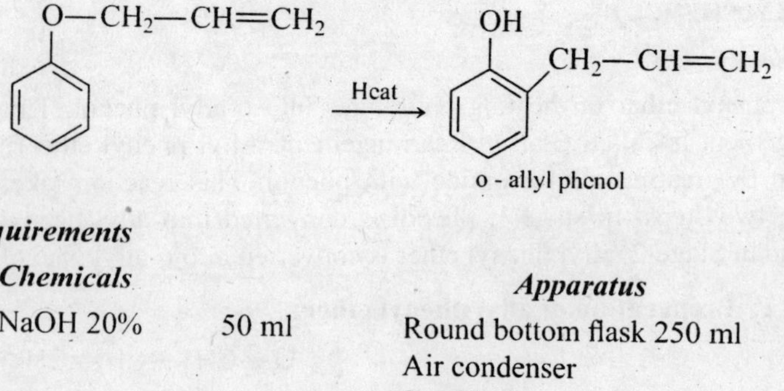

o - allyl phenol

Requirements

Chemicals

Chemicals		Apparatus	
NaOH 20%	50 ml	Round bottom flask 250 ml	1
		Air condenser	1
		Sand bath	1
		Abbe refractometer	1

Procedure

The allylphenyl ether is heated in a flask fitted with a long air condenser over a sand bath and kept gently refluxing. The progress of the reaction is followed by measuring the refractive index with the Abbe Refractometer at frequent intervals. The reaction may be completed in 5-6 hours. When the refractive index has risen to 1.5433 from its initial value of 0.9831, the transformation may be taken as complete.

The distilled product is dissolved in twice its volume of 20% sodium hydroxide solution and the alkaline solution is extracted twice with 30 ml of petroleum ether. The alkaline solution is then acidified with dilute hydrochloric acid and extracted repeatedly with ether. The ether extract is dried over anhydrous calcium chloride. After removal of ether the 2-allylphenol is distilled under reduced pressure.

(Expected yield: 3 g; Boiling point. 104–105°C).

(08) FRIES REARRANGEMENT (2-HYDROXY-5-METHYLBENZOPHENONE)

Principle

Substituted phenols on benzoylation give benzoyl phenol derivatives which undergo the Fries rearrangment give hydroxybenzophenone. This preparation involves two stages. In stage-1, p-cresol is converted into p-cresyl benzoate. In Stage-2, p-cresyl benzoate is converted into 2-hydroxy-5-methylbenzophenone by Fries rearrangement.

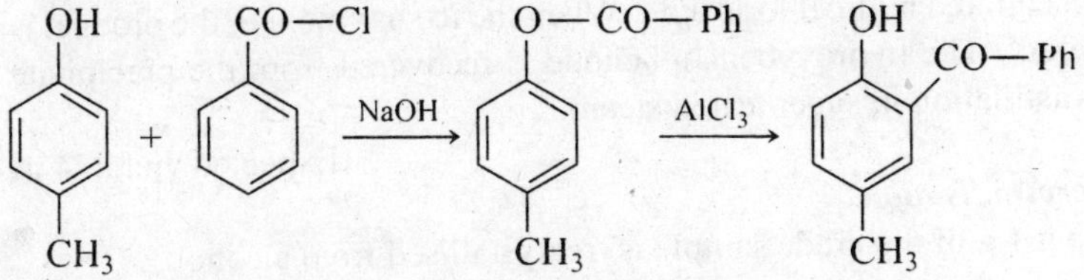

Stage –1: Preparation of p-cresyl benzoate

Requirements

Chemicals			*Apparatus*	
p-cresol	5 g		Round bottom flask	1
Benzoyl chloride	5 g		Reflux condenser	1
NaOH (2N)		20 ml	Beaker 250 ml, Separating funnel	
Anhy. $AlCl_3$		4 g	Watch glass, glass rod, Funnel	
Conc. HCl		20 ml	Water bath	1

Procedure

The p-cresol,100 ml of water, benzoyl chloride are taken in a round bottom flask. To this mixture, enough sodium hydroxide solution (2N) is added till strongly alkaline and the flask shaken vigorously for about half an hour after suitably corking it. Heat is evolved and the flask is cooled under the tap. The oily product is poured into ice water and rubbed with a glass rod to induce crystallisation. The solid is filtered. washed with plenty of water and dried.

(Expected yield: 6 g)

Recrystallisation

About 1 g of the crude sample is crystallised from 50% alcohol. The crystals are filtered at the pump and dried in a vacuum desiccator over concentrated sulphuric acid.

(Melting point: 71°C)

Stage – 2: Preparation of 2-Hydroxy-5-methylbenzophenone

The p-Cresyl benzoate and anhydrous aluminium chloride are introduced into a bolt-head flask fitted with a two holed cork carrying a thermometer and an air condenser. A calcium chloride guard-tube is attached to the condenser. The reactants are mixed by shaking the flask and the flask is then heated in an oil-bath kept at 90°C. The temperature of the bath is raised rapidly to 120°C and then slowly to 140°C at which it is kept for about 15 minutes. The thermometer is removed and when the flask is cooled, the contents are poured into of ice and 20 ml of concentrated hydrochloric acid. When the ice has melted, the product is filtered. The hydroxybenzophenone is recovered from the precipitate by distillation in superheated steam.

(Expected yield: 4 g)

Recrystallisation

About 1 g of the crude sample is recrystallised from alcohol.

(Melting point: 83–84°C)

(09) ARNDT-EISTERT SYNTHESIS (α-NAPHTHAYL ACETIC ACID)

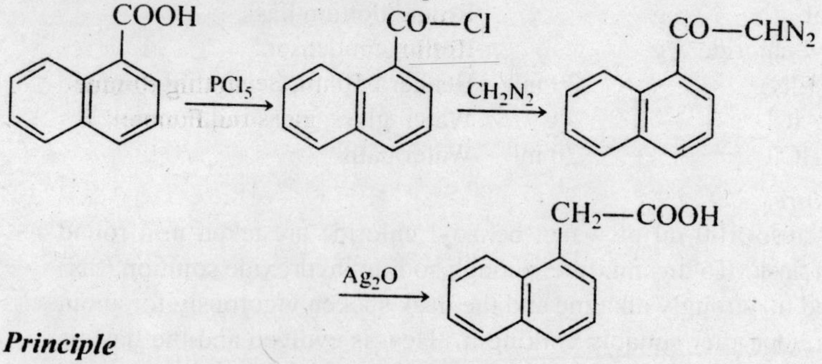

Principle

α-Naphthoic acid can be converted into α-Naphthylacetic acid in a reaction called Arndt-Eistert Synthesis. This reaction takes place in 2 stages. In stage-1, α-naphthoic acid is first converted into α-naphthoyl chloride by treating with phosphorus pentachloride. In stage-2, this product is converted into α-naphthoyl acetic acid by treating with diazomethane and silver oxide along with sodium thiosulphate.

Stage – 1: Preparation of α-Naphthoyl chloride

Requirements

Chemicals		*Apparatus*
α-Naphthoic acid	10 g	Round bottom flask
Nitrosomethyl urea	15 g	Reflux condenser
KOH (40%)	55 ml	Beaker 250 ml, Separating funnel
Ether	300 ml	Watch glass, glass rod, Funnel
PCl_5	15 g	Water bath

Procedure

This preparation requires diazomethane solution which can be prepared as follows: A mixture of 200 ml of ether and 55 ml of 40% aqueous potassium hydroxide is cooled to 5°C. To this nitrosomethyl urea is added with continuous cooling and efficient stirring. The deep yellow ethereal solution of diazomethane is separated by means of a separating funnel and dried over pure potassium hydroxide pellets for an hour. The solution contains about 5 g of diazomethane.

A mixture of 1-napthoic acid and phosphorous pentachoride is placed in a round bottom flask and heated on a water-bath until no more hydrogen chloride is evolved. The phosphorous oxychloride is then removed by distillation under reduced pressure. The residue is then distilled using an oil pump. The α-Naphthoyl chloride is collected at 168°C.

(Expected yield: 8 g)

Stage – 2: Preparation of α-Naphthaylacetic acid

A solution of α-Naphthoyl chloride in 25 ml of dry ether is added dropwise to an ethereal solution of diazomethane cooled in ice without allowing the temperature to rise above 10°C. The reaction may take about six hours and the ether is removed in vacuum. The residue is α-naphthyl diazoketone. A solution of the diazoketone in 25 ml of dioxane is added to a solution of 2 g silver oxide and 3 g sodium thiosulphate in 100 ml water and heated on a water bath at 50–60°C with stirring. The stirring is continued, till nitrogen evolution ceases. The solution is then cooled, made acidic with dilute nitric acid. The precipitated acid is filtered off at the pump, washed with water and dried.

(Expected yield: 1g)

Recrystallisation

A small portion of the crude sample is recrystallised from boiling water.

(Melting point: 130°C)

(10) GRIGNARD REACTION (TRIPHENYL CARBINOL)

Principle

Phenyl magnesium bromide on treatment with ethyl benzoate, followed by hydrolysis gives triphenyl carbinol. This reaction is called Grignard reaction. The preparation of Grignard reagent should be done in dry medium because even a few drops of water may hydrolyse the reagent.

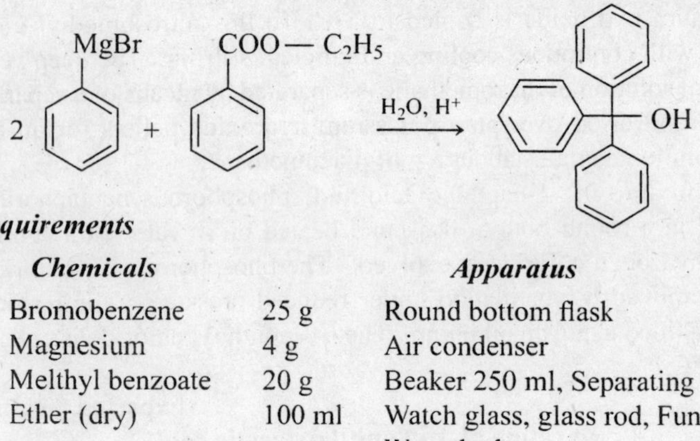

Requirements

Chemicals		*Apparatus*	
Bromobenzene	25 g	Round bottom flask	1
Magnesium	4 g	Air condenser	1
Melthyl benzoate	20 g	Beaker 250 ml, Separating funnel	
Ether (dry)	100 ml	Watch glass, glass rod, Funnel	
		Water bath	1

Procedure

Grignard reagent is prepared as follows: Clean magnesium turnings of ribbon in small pieces are washed with ether, dried at 110–120°C in an air-oven and introduced into a dry round-bottom flask, fitted with a one-holed cork carrying an upright dry condenser protected by a calcium chloride tube and containing pure dry bromobenzene dissolved in ether.

The flask warmed on a water-bath until the magnesium has completely dissolved. Freshly distilled methyl benzoate dissolved in ether is added to the cold solution slowly with constant shaking. The mixture is boiled under reflux for an hour, cooled and acidified with dilute hydrochloric acid till acidic, after adding crushed ice. The ether is removed by distillation and the residue stream distilled to remove unreacted bromobenzene as well as any diphenyl formed.

The triphenyl carbinol which remains over, is filtered washed with water and dried

(Expected yield: 10 g)

Recrystallisation

A small portion of the sample is recrystallised from benzene. Triphenyl carbinol forms as colourless needles.

(Melting point: 162.5°C)

B-II : TWO-STAGE PREPARATIONS

(01) p–Bromoacetanilide from Aniline

Stage 1 : *Acetanilide from aniline*

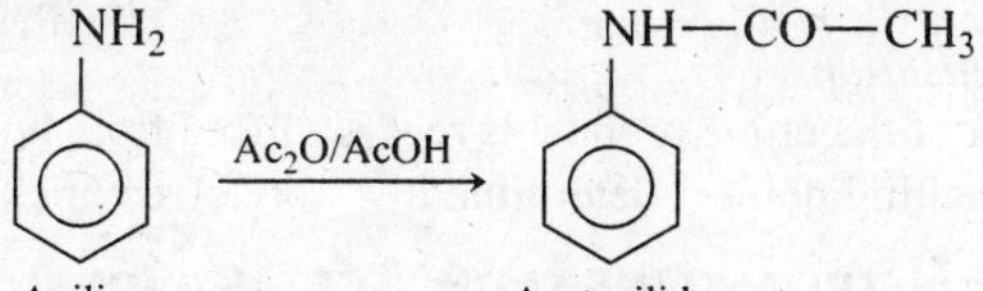

Aniline — Acetanilide

Requirements

Chemicals		*Apparatus*	
Aniline (freshly distilled)	5 ml	Round-bottom flask (250 ml)	1
Acetic anhydride	7 ml	Reflux-condenser	1
Glacial acetic acid	7 ml	Wire-gauze	1
Rectified spirit	10 ml	Glass beaker (250 ml)	1
		Test tube (10 ml)	1

Procedure

Same as page-36 (*see*, single-stage section), using the above quantities.

(Expected yield : 7 g)

Recrystallisation

About 1 g of the crude product is recrystallised from hot water, and the melting point is determined.

(Correct melting point : 114°C)

Stage 2 : *p-Bromoacetanilide from acetanilide*

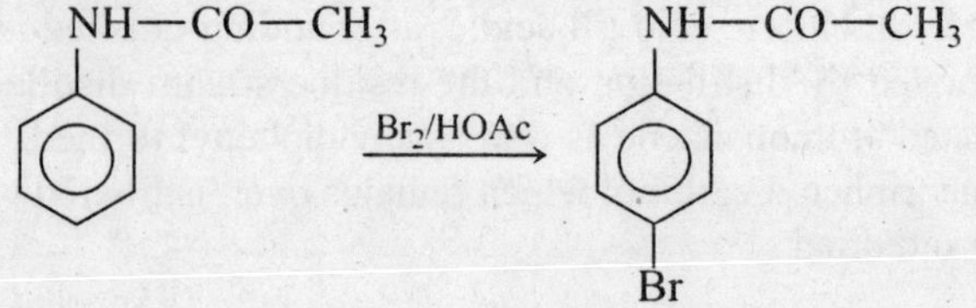

Requirements

Chemicals		Apparatus	
Acetanilide	5 g.	Conical flask (250 ml)	1
Bromine	7 ml	Cork	1
Glacial acetic acid	40 ml	Boiling test-tube	1
		Dropper	1
		Glass rod	1

Procedure

Same as page-42 (*see*, single-stage section), using the above quantities.

(Expected yield : 7 g.)

Recrystallisation

About 1 g of the crude product is recrystallised from hot rectified spirit, and the melting point is determined. (Correct melting point : 7 g.)

(02) p-Nitroaniline from Acetanilide

Stage 1 : *p – nitroacetanilide from acetanilide*

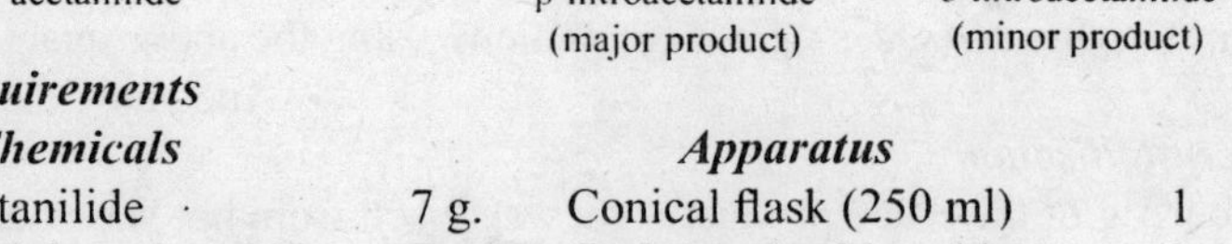

acetanilide — p-nitroacetanilide (major product) — o-nitroacetanilide (minor product)

Requirements

Chemicals		Apparatus	
Acetanilide	7 g.	Conical flask (250 ml)	1
Glacial acetic acid	7 ml	Cork (preferably rubber)	1
Fuming nitric acid	3 ml	Beaker (250 ml)	1
Con. sulphuric acid	15 ml	Boiling test-tube	1
Sodium carbonate	25 g.		
Rectified spirit	10 ml		

Procedure

Acetanilide (7 g) is dissolved in 7 ml glacial acetic acid in a conical flask. Con. sulphuric acid (15 ml) is added very slowly, with constant shaking. The flask is cooled in a freezing mixture of crushed ice and crude common salt. Then, fuming nitric acid (3 ml) is added, very slowly with gentle shaking, to the flask so that the temperature does not rise above 20ºC. The contents of the flask are shaken vigorouly (using a mechanical shaker, if available) for about an hour. Then, it is poured into cold water in a thin stream, slowly with shaking. The product is allowed to stand for an hour, filtered off at the pump, and then washed with water.

The ortho-isomer obtained as the minor product is removed by stirring the above product with water, and adding sufficient amount of sodium carbonate till the liquid portion becomes alkaline to litmus. The mixture is then boiled for a few minutes. The ortho-product undergoes hydrolysis and goes into solution, but the para-product is not affected and remains in the solid state. This solid is filtered off, washed wi water, dried and the yield is noted.

(Expected yield : 8

Recrystallisation

About 1 g of the crude sample is recrystallised from hot rectified spirit. The melting point is determined.

(Correct melting point : 214ºC)

Stage 2 : *p-nitroaniline from p-nitroacetanilide*

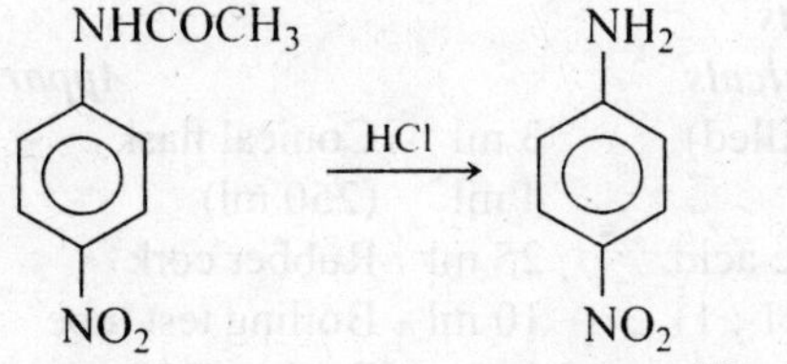

p-nitro acetanilide

Requirements

Chemicals		***Apparatus***	
p-nitroacetanilide	6 g	Conical flask (250 ml)	1
Con. HCl	15 ml	Round-bottom flask (250 ml)	1
Aq. ammonia (4 N)	30 ml	Air-condenser	1
		Beaker (250 ml)	1

Procedure

The *p*-nitroacetanilide (6 g) is mixed with con. HCl (15 ml) in a round-bottom flask, fitted with an air-condenser, and refluxed for about 40 minutes (or till the liquid remains clear on dilution with water). The *p*-nitroaniline is liberated on neutralising the reaction mixture with an excess of aqueous ammonia. The contents are cooled. The yellow precipitate that separates out is filtered off at the pump, washed with cold water, dried and the yield is noted.

(Expected yield : 6 g.)

Recrystallisation

About 1 g of the crude sample is recrystallised from boiling water and the melting point is determined.

(Correct melting point : 148°C.)

(03) 1, 3, 5-TRIBROMOBENZENE FROM ANILINE

STAGE 1 : *2, 4, 6-Tribromoaniline from Aniline*

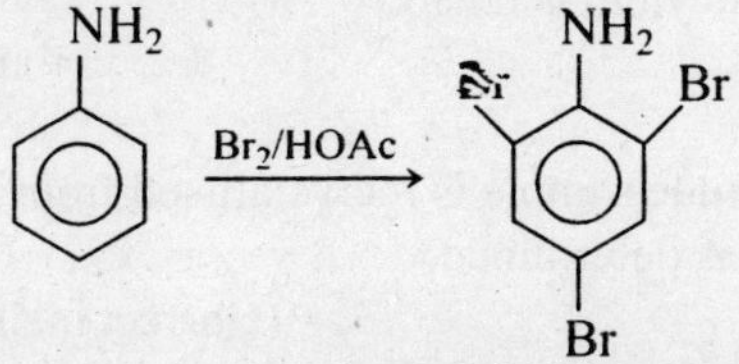

Syno - 2,4,6 tribromo aniline

Requirements

Chemicals		*Apparatus*	
Aniline (distilled)	3 ml	Conical flask	1
Bromine	4 ml	(250 ml)	1
Glacial acetic acid	25 ml	Rubber cork	1
Aq. ethanol (1 : 1)	10 ml	Boiling test tube	1
Dropper	1	Beaker (250 ml)	1

Procedure

Same as page-41 (*see*, single-stage section), using the above quantities.

(Expected yield : 8 g)

Recrystallisation

About 1 g of the crude sample is recrystallised from 1 : 1 aq. alcohol. and the melting point is determined.

(Correct melting point : 120°C.)

Stage 2 : *Sym-tribromobenzene from Tribromoaniline*

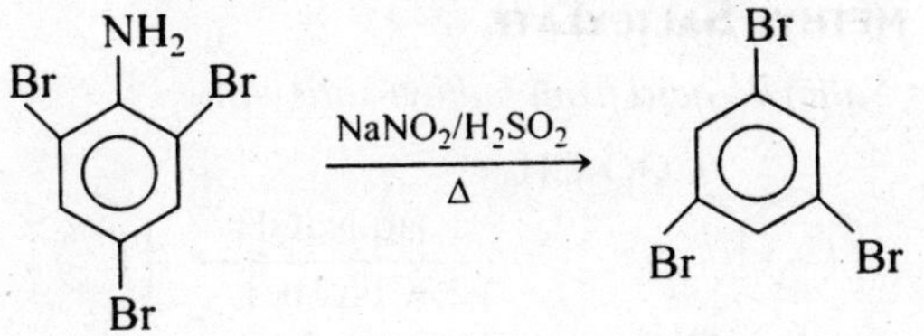

2,4,6-tribromoaniline 1,3,5-tribromobenzene

Recrystallisation

About 1 g of the sample is recrystallised from hot rectified spirit, and the melting point is determined.

(Correct melting point : 122°C)

Requirements

Chemicals		*Apparatus*	
2,4,6 - Tribromoaniline	5 g.	Round-bottom flask (250 ml)	1
Aq. NaOH solution	40 ml	Reflux-condenser	1
Rectified spirit	7 ml	Beaker (250 ml)	1
Con. sulphuric acid	2 ml	Boiling test tube	1
Sodium nitrite	2 g.		
Rectified spirit	10 ml		

Procedure

2, 4, 6-Tribromoaniline (5 g), rectified spirit (30 ml) and benzene (7 ml) are mixed in a round-bottom flask, fitted with a reflux condenser, and heated on a water-bath, till the solution starts boiling. About 1 g sodium nitrite is added in one lot through the top of the condenser, and the mixture is shaken vigorously. It is then heated on the boiling water-bath, till no more gas is evolved. The solution is then cooled, preferably in ice. The solid that separates out is filtered off at the pump, washed thoroughly with water, dried and the yield is noted.

(Expected yield : 4 g.)

Recrystallisation

About 1 g of the sample is recrystallised from hot rectified spirit, and the melting point is determined.

(Correct melting point : 122°C.)

(04) ACETYL SALICYLIC ACID (ASPIRIN) FROM METHYL SALICYLATE

Stage 1 : *Salicylic acid from methyl salicylate.*

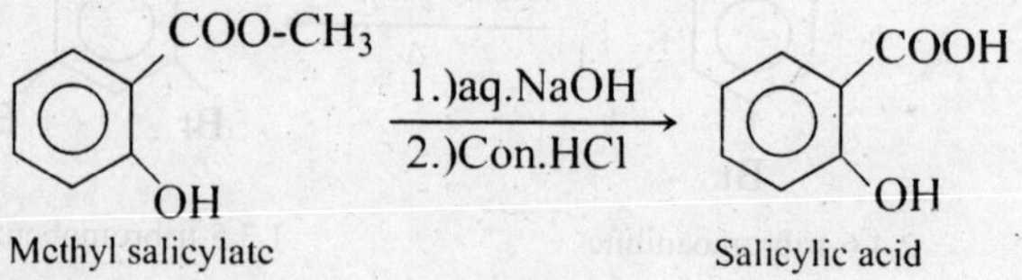

Requirements

Chemicals		*Apparatus*	
Methyl salicylate	5 ml	Round-bottom flask (250 ml)	1
Aq. NaOH solution	40 ml	Reflux-condenser	1
Con. HCl	20 ml	Beaker (250 ml)	1
		Glass rod	1
		Wire-gauze	1

Procedure

Same as page-34 (*see*, single-stage section), using the above quantities.

(Expected yield: 6 g)

Recrystallisation

About 1 g of the crude sample is recrystallised from boiling water, and the melting point is determined.

(Correct melting point: 159ºC)

Stage 2 : *Acetyl salicylic acid from salicylic acid*

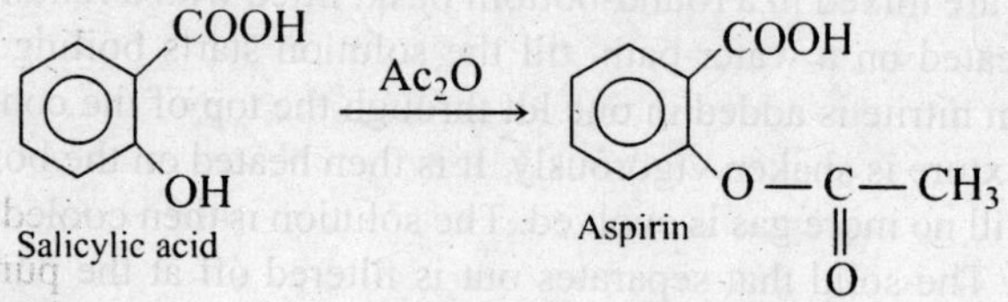

Requirements

Chemicals		*Apparatus*	
Salicylic acid	4 g.	Conical flask (100 ml)	1
Acetic anhydride	6 ml	Water-bath	1
Con. sulphuric acid	4 dps.	Glass beaker (250 ml)	1

Procedure

Same as page-37 (*see*, single-stage section), using the above quantities.

(Expected yield: 5 g.)

Recrystallisation

Same recrystallisation procedure given in page-38.

(05) Benzilic Acid from Benzoin

Stage 1: Benzil from Benzoin

$$C_6H_5COCH(OH)C_6H_5 \xrightarrow{HNO_3} C_6H_5COCOC_6H_5$$

Requirements

Chemicals		*Apparatus*	
Benzoin	7 g.	Round-bottom flask (250 ml)	1
Con. nitric acid	15 ml	Air-condenser	1
Rectified spirit	20 ml	Beaker (250 ml)	1

Procedure

Benzoin (7 g) and con. HNO_3 (15 ml) are mixed in a 250 ml round-bottom flask, and heated on a water-bath fitted with an air-condenser. The mixture is occasionally shaken well. The contents are heated for two hours (students are advised to conduct this experiment in fume cupboard, as the reaction is very vigorous with copious evolution of nitrogen dioxide fumes). After heating, the product is poured into cold water in a 250 ml beaker with vigorous stirring. The crystalline solid that separates out is filtered off at the pump, washed with cold water, dried and the yield is noted.

(Expected yield: 6 g.)

Recrystallisation

About 0.5 g of the crude benzil sample is recrystallised from hot rectified spirit, and the melting point is determined.

(Correct melting point: 95°C)

Stage 2 : *Benzilic acid from Benzil*

Requirements

Chemicals		*Apparatus*	
Benzil	5 g	Conical flask (250 ml)	1
Caustic potash	5 g/20 ml	Round-bottom flask (250 ml)	1
Rectified spirit	15 ml	Reflux condenser	1
Dil. H_2SO_4		Water - bath	1
		Beaker (250 ml)	1

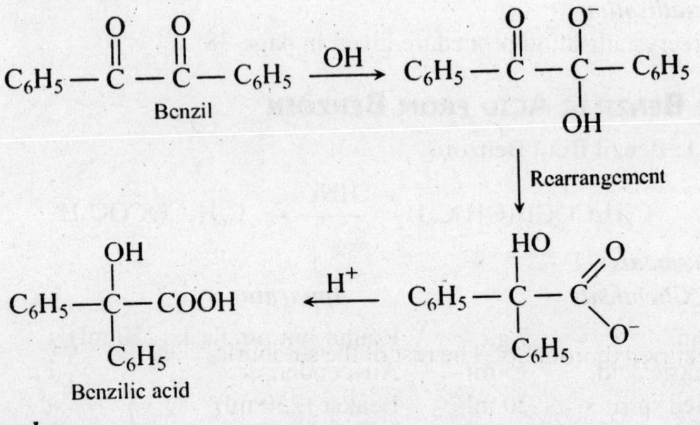

Procedure

Benzil (5 g) is dissolved in rectified spirit (15 ml) in a round-bottom flask, and the caustic potash soln. (20 ml) is added. The mixture is refluxed for 20 minutes on boiling water-bath. The contents are then poured into a 250 ml beaker, and cooled first in tap-water and then in crushed ice. The potassium salt of benzilic acid that separates out is filtered off at the pump, and washed with rectified spirit. Then about 100 ml of water is added, and the solution is boiled. It is then acidified with hot dil. H_2SO_4 till the solution becomes acidic to congo-red paper. The solid that crystallises out is filtered off, washed with water, dried and the yield is noted.

(Expected yield: 5 g.)

Recrystallisation

About 1 g of the crude sample is recrystallised from hot water, and the melting point is determined.

(Correct melting point: 150ºC)

(06) m-NITROANILINE FROM NITROBENZENE

STAGE 1 : *m-Dinitrobenzene from Nitrobenzene*

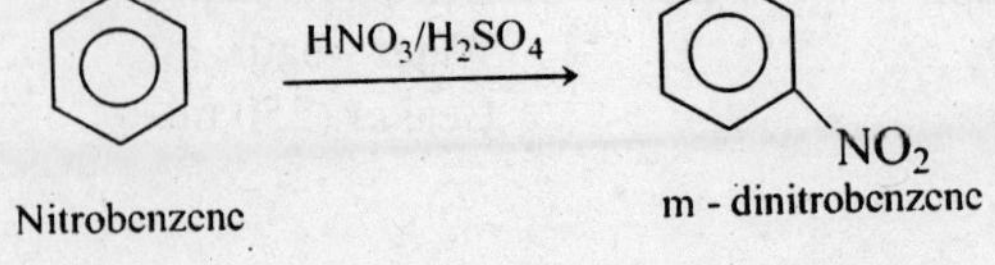

Requirements

Chemicals		***Apparatus***	
Nitrobenzene	7 ml	Round-bottom flask (250 ml)	1
Con. sulphuric acid	8 ml	Air-condenser	1
Fuming nitric acid	12 ml	Water-bath	1
Rectified spirit	10 ml	Beaker (250 ml)	1
		Boiling test tube	1

Procedure

Same as page-44 (see, single-stage section), using the above quantities.

(Expected yield: 7 g)

Recrystallisation

Same recrystallisation procedure given in page-45 (single-stage section).

(Correct melting point: 90°C)

Stage 2 : *m-Nitroaniline from m-Dinitrobenzene.*

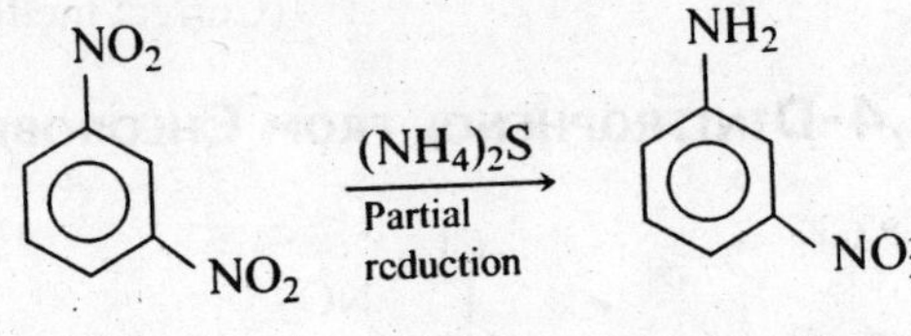

m - dinitrobenzene m - nitroaniline

Requirements

Chemicals		***Apparatus***	
Acetanilide	7 g.	Conical flask (250 ml)	1
m-Dinitrobenzene	6 g.	Round-bottom flask (250 ml)	1
Rectified spirit	25 ml	Reflux condenser	1
Liquor ammonia	3ml	Beaker (250 ml)	1
Hydrogen sulphide			
Dil HCl	20 ml		

Procedure

The crude *m*-dinitrobenzene (6 g) is dissolved in hot rectified spirit (25 ml) in a round-bottom flask. The contents are cooled well by immersing the flask in cold water. H_2S gas is then passed into the solution for about an hour. The flask is fitted with a reflux condenser, and heated for about 20 minutes on a boiling water-bath. Two drops of the liquid from the flask are removed and poured into water. If a solid appears the reaction is considered as complete. (If no solid appears, the mixture is cooled well, and H_2S gas is passed for another 20 minutes, refluxed for 10 minutes, tested again.) The reaction mixtures in the flask is cooled well under the tap, and poured into cold water with constant stirring. (The solid *m*-nitroaniline that separates out is filtered off at the pump, and washed with water.) To get a further yield of *m*-nitroaniline, the mother liquor is mixed with 20 ml dil. HCl and concentrated to half the volume by boiling. The concentrated liquid is then neutralised with an excess of ammonia, when solid *m*-nitroaniline separates out. The solid that separates out is filtered off and washed with water and dried. The total yield is noted.

(Expected yield: 4 g.)

Recrystallisation

About 1 g of the crude sample is recrystallised from boiling water and the melting point is determined.

(Correct melting point: 112ºC)

(07) 2,4-Dinitrophenol from Chlorobenzene

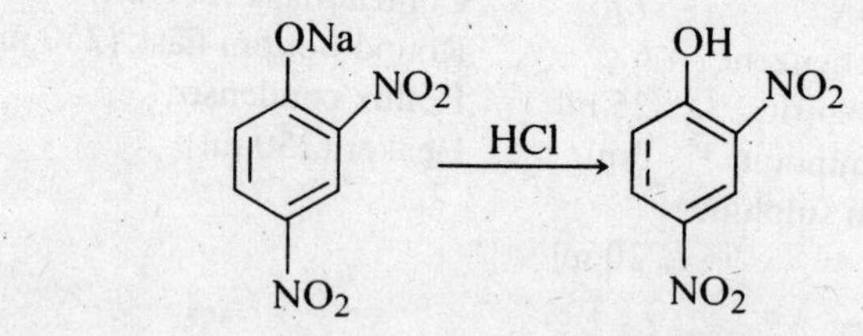

2,4 - Dinitrophenol

Principle:

Chlorobenzene can be converted into 2,4-dinitrophenol in two stages. In stage-1, chlorobenzene is converted into 2,4-dinitrochlorobenzene by nitration using a nitrating mixture of conc. sulphuric acid and conc. nitric acid. In stage 2, 2,4-dinitrochlorobenzene is converted into 2,4-dinitrophenol using a base by nucleophilic substitution reaction.

(a) 2, 4-Dinitrochlorobenzene

Requirements

Chemicals		*Apparatus*	
Chlorobenzene	15 ml	Round bottom bolt-head flask	1
Conc.Sulphuric Acid	35 ml	Dropping funnel	1
Conc. Nitric acid	20 ml	Reflux condenser	1
		Water bath, stirrer, beaker	1

Procedure

The chlorobenzene is mixed with half the quantity of sulphuric acid in a measured into a 250 ml round bottom bolt-head flask. The mixture is warmed on a water both and stirred thoroughly. The rest of the sulphuric acid is added little by little to the nitric acid with cooling under tap. The acid mixture is transferred to a dropping funnel and added drop by drop so that the temperature remains below 100°C. After the addition is over, the reaction mixture is stirred for an hour on a water bath. The reaction is completed by heating the reaction mixture for another hour at 120°C in an oil bath, using an air condenser. The liquid is partly cooled and poured on to crushed ice contained in a beaker and stirred. The solid product is filtered, washed well with cold water and pressed and dried.

(Expected yield: 16 g)

Recrystallisation

About 1 g of the crude solid is purified by crystallization from rectified spirit.

(Meling point: 53°C)

Caution

1. Contact of the substance with the skin may cause dermatitis
2. As the vapours are poisonous, it is advisable to use gloves in this preparation.

(b) 2, 4-DINITROPHENOL FROM 2,4-DINITROCHLOROBENZENE

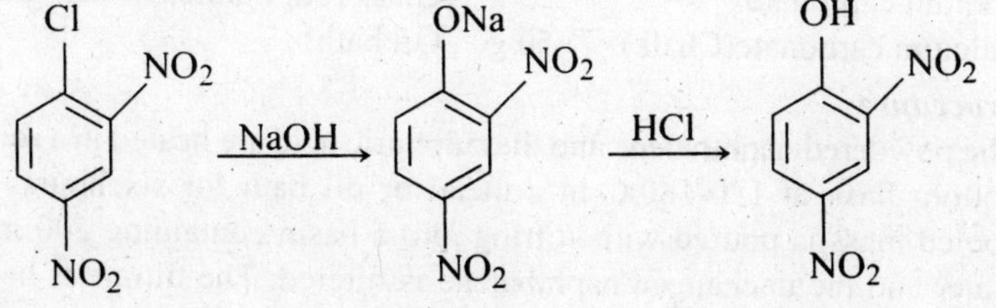

Requirements

Chemicals		*Apparatus*	
2,4-dinitrochlorobenzene	10 g	Round bottom flask	1
Sodium carbonate (anhydrous)	15 g	Reflux condenser	1
Or Sodium hydroxide	10 g	Funnel, glass rod.	1
Dilute HCl	30 ml		

Procedure

The sodium carbonate or hydroxide is dissolved in 100 ml of water contained in a round bottom flask. The 2,4-dinitrochlorobenzene is added and the mixture heated using a water-cooled condenser till the oily layer disappears, and the liquid becomes clear. It may take about 2-3 hours. The clear liquid is filtered hot using a Buchner funnel and the filtrate acidified with hydrochloric acid. On cooling, the dinitrophenol separates. It is filtered, washed with cold water and dried.

(Expected yield: 8 g)

Recrystallisation

The crude crystals of dinitrophenol is recrystallised from minimum amount of hot water.

(Melting point: 112°C)

(08) β-NAPHTHOL FROM NAPHTHALENE

Stage: 1 *Sodium β-naphthalene Sulphonate from Naphthalene*

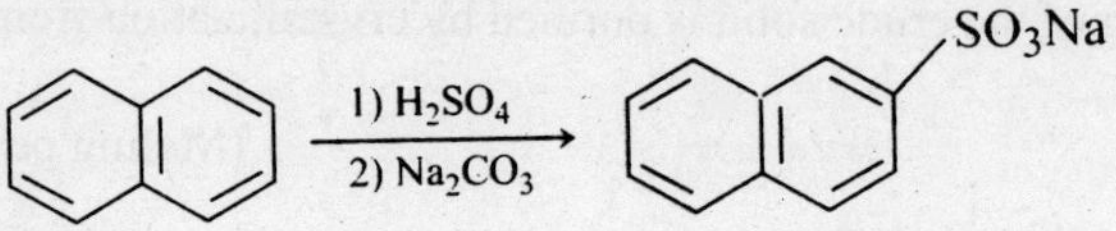

Requirements

Chemicals		*Apparatus*	
Naphthalene	20 g	Round bottom flask 250 ml	1
Concentrated sulphuric acid	25 ml	Beaker 500 ml	1
Sodium carbonate	25 g	Glass rod, Funnel, Watch glass	
Calcium carbonate(Chalk)	50 g	Oil bath	1

Procedure

The powdered naphthalene and the sulphuric acid are heated in a round-bottom flask at 170-180°C in a metal or oil bath for six hours. The cooled mass is poured with stirring into a basin containing 500 ml of water and the unchanged naphthalene is filtered. The filtrate is heated to boiling and neutralised by the addition of chalk or slaked lime in the form of a thick cream.

The precipitated calcium sulphate is filtered hot. The residue washed with 50 ml of water. The combined filtrate and washings are concentrated by evaporation until a sample crystallizes on cooling. Saturated sodium carbonate solution is added to the hot liquid till no further precipitation occurs. The precipitated chalk is filtered and the filtrate is evaporated almost to dryness, when the sodium salt crystalises out. The salt is filtered, washed with water and dried on a water-bath.

(Expected Yield: 20 g)

(b) Stage 2: β-*naphthol from Sodium* β-*naphthalene Sulphonate*

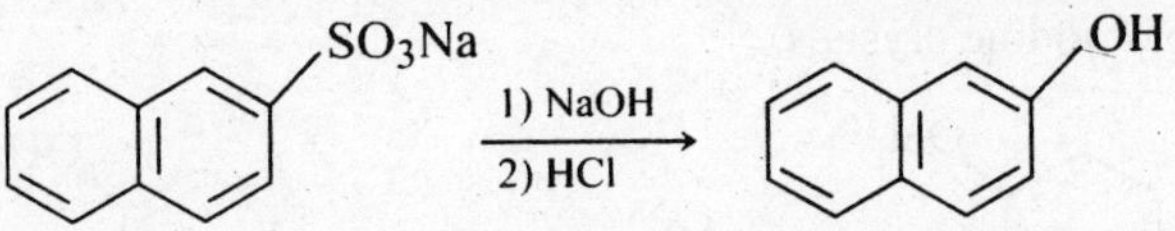

Principle

Sodium β-Naphthalene sulphonate on hydrolysis in basic medium gives sodium napthoxide which on acidification gives β-Naphthol.

Requirements

Chemicals		*Apparatus*
Sodium β-Naphthalene sulphonate	20 g	Nickel crucible
Caustic soda	60 g	Thermometer
Water	100 ml	Beaker 250 ml, glass rod

Procedure

The caustic soda and water are heated in a nickel crucible or basin carefully till a temperature of about 280°C is reached. The molten mass is stirred by a closed tube containing a little mercury into which the thermometer is dipped. The powdered sodium salt is added a little at a time and the temperature gradually raised to 300°C. The mass shows a tendency to froth up and assumes a light yellow coloured.

The mass is heated at this temperature for about 10 minutes. The completion of the reaction is shown by the separation of two layers and the darkening of the colour. The melt is poured into a clean tray, pulverised when cold, dissolved in the minimum amount of water and acidified with (1:1) hydrochloric acid. The precipitate of β-naphthol is filtered, washed with water and dried.

Recrystallisation

About 1 g of the crude β-naphthol sample is recrystallised from hot water. It crystallises as colourless leaflets.

(Expected yield: 10 g; melting point : 122°C)

(09) Coumarin from Phenol

Principle

The preparation of coumarin involves two stages. In stage-1, phenol is converted into salicylaldehyde by Reimer-Tiemann reaction using chloroform and alkali. In stage-2, salicylaldehyde is converted into coumarin by treating with acetic anhydride and sodium acetate in the presence of iodine crystals.

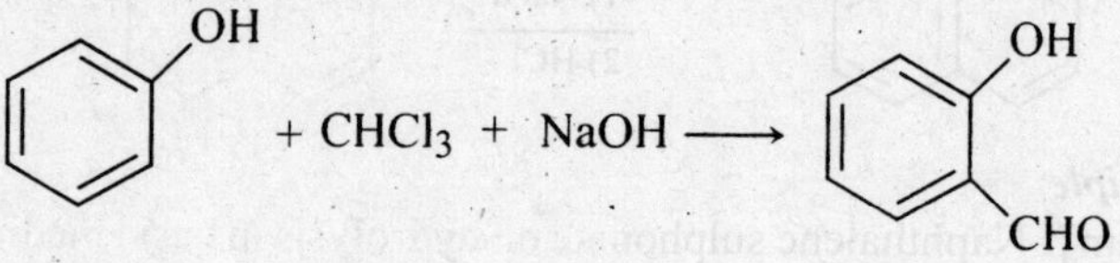

Stage 1: Preparation of Salicylaldehyde

Requirements

Chemicals		Apparatus
Phenol	10 g	Round-bottom flask 250 ml
Chloroform	30 g	Reflux condenser
Caustic soda/100 ml water	20 g	Water bath, Beaker, funnel, glass rod.

Procedure

The phenol is dissolved in caustic soda solution in a round-bottom flask. The flask is fitted with a reflux condenser. The chloroform is added through the condenser in lots of 5 ml and flask is well shaken after each addition. Throughout the addition, the temperature is kept at 60°-65°C by immersion in a hot water-bath. The colour changes from yellow through violet to deep red.

The mixture is then refluxed for half an hour till the separation of a yellow semi solid mass takes place. The unreacted chloroform is removed by distillation on a water bath, the residual liquid is diluted with water, carefully neutralised with dilute sulphuric acid and steam distilled till no more oily drops pass over. The distillate consists of a mixture of salicylaldehyde and phenol and these two can be separated through the formation of the bisulphite compound of the aldehyde. The distillate is extracted with ether and the ether solution shaken with a freshly prepared saturated solution of sodium bisulphite continuously till no more crystals of the bisulphite compound are obtained. The white crystals obtained are filtered, washed with alcohol to remove traces of phenol and decomposed by heating with dilute sulphuric acid on water bath. The aldehyde is extracted with ether, dried over calcium chloride and after removal of ether distilled. It is obtained as a colourless fragrant oil.

(Expected yield: 2 g; Boiling point: 196.5°C)

Stage 2: Preparation of Coumarin

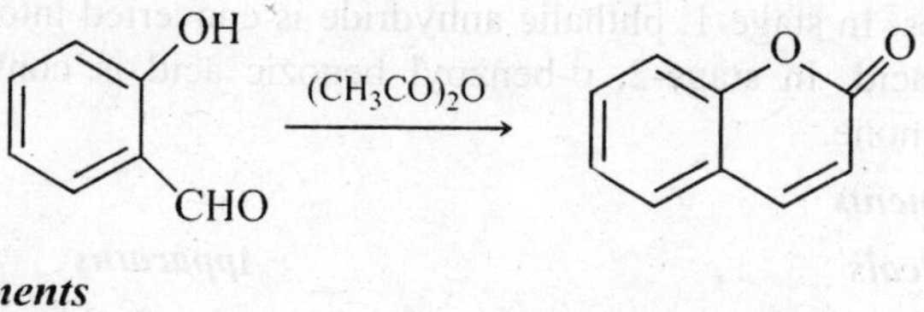

Requirements

Chemicals		***Apparatus***	
Salicylaldehyde	2 g	Round-bottom flask 250 ml	1
Acetic anhydride	5 g	Air condenser	1
Sodium acetate	5 g	Oil bath, Thermometer, Beaker 250 ml, Glass rod,	
Iodine	0.2 g	Separating Funnel.	1

Procedure

A mixture of salicylaldehyde, acetic anhydride, sodium acetate and iodine is taken in a round bottom flask, fitted with an air condenser and heated in an oil bath for about 2 hours at 120°C. The product is then transferred in to a separating funnel, mixed with water and extracted with ether. The ether solution is separated and evaporated where crystals of coumarin appears.

(Expected yield: 2 g; Melting point: 67°C)

(10) ANTHRAQUINONE FROM PHTHALIC ANHYDRIDE

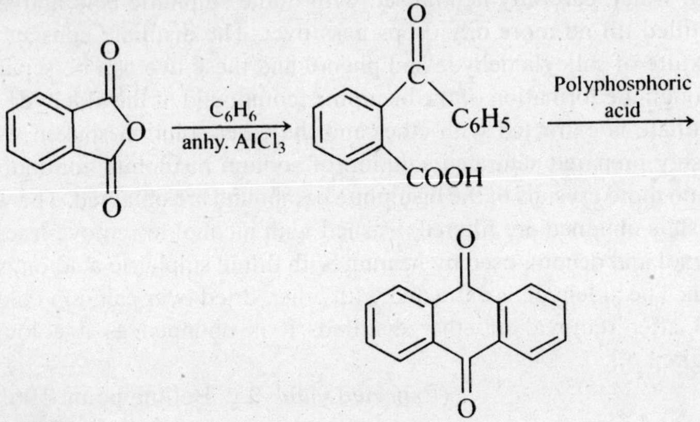

Principle

The preparation of anthraquinone from phthalic anhydride involves two stages. In stage-1, phthalic anhydride is converted into o-benzoyl benzoic acid. In stage-2, o-benzoyl benozic acid is converted into anthraquinone.

Requirements

Chemicals / ***Apparatus***

Phthalic anhydride	3g	Round bottom flask	1
Phosphorus pentoxide	15g	Reflux condenser	1
Ortho phosphoric acid	20g	Water bath, Beaker 250 ml	1

Procedure

(a) Stage-1: Preparation of o-benzoylbenzoic acid from phthalic anhydride (refer procedure from page - 90).

(b) Stage-2: Preparation of anthroquinone from o-benzoylbenzoic acid.

Polyphosphoric acid is prepared by dissolving 15 g of phoshorus pentoxide in 20 g of orthophosphoric acid. A mixture of polyphosphoric acid and o-benzoylbenzoic acid is taken in a round bottom flask, fitted with a reflux condenser. The reaction mixture is heated on a steam bath for about 2 hours. The reaction mixture is then cooled and poured into ice cold water and stirred well. The solid separated out is filtered off at the pump washed with water and dried.

(Expected yield:

Recrystallisation

About 1 g of the crude sample is recrystallised from hot glacial acetic acid in the presence of charcoal.

(Melting point: 286°C)

B–III Three-Stage Preparations:

(01) 1-Bromo-2-bromomethyl naphthalene from 2-methyl naphthalene

Principle

This preparation involves three stages. In stage-1, 2-Methyl naphthalene is converted into 1-bromo-2-methylnaphthalene on bromination using bromine in carbon tetra chloride. In stage-2, N-bromosuccinimide is prepared from succinimide on bromination. In Stage-3, the methyl group present in the position –2 is brominated to 2-bromomethyl thus converted 2-methylnaphthalene into 1-bromo-2-bromomethyl naphthalene.

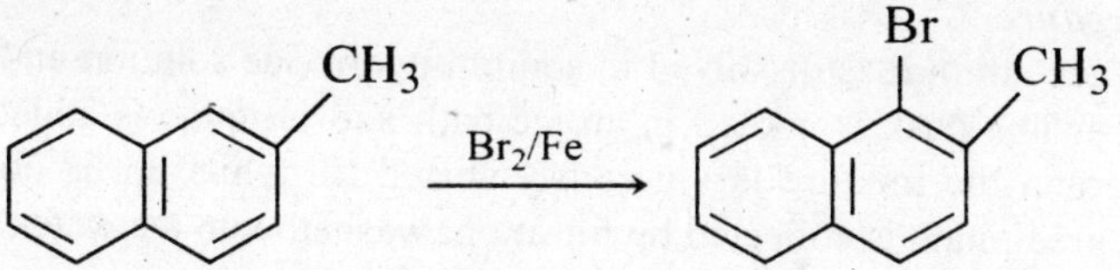

Stage 1: Preparation of 1-Bromo-2-methyl naphthalene

Requirements

Chemicals		*Apparatus*	
2-Methyl Naphthalene	14 g	Conical flask	1
Bromine	16 g	Thermometer	1
Carbon tetra chloride	60 ml	Stirrer	1
Iodine crystal/iron powder		Watch glass	1
NaOH (10% solution)	30 ml	Glass rod.	1

Procedure

To 2-methylnaphthalene in carbon tetrachloride, a crystal of iodine and a pinch of iron powder are added. The mixture is cooled to 0°C and treated drop wise with a solution of bromine in carbon tetrachloride(about 30 ml). The reaction mixture is stirred well during the addition and the flask is covered with black paper to prevent the exposure of light. The temperature is not allowed to rise above 5°C. The mixture is left over night and is then washed successively with 10% sodium hydroxide solution and water. After drying over calcium chloride, the solvent is removed and the residual liquid distilled.

(Expected yield: 18 g)

Stage 2: Preparation of N-Bromosuccinimide

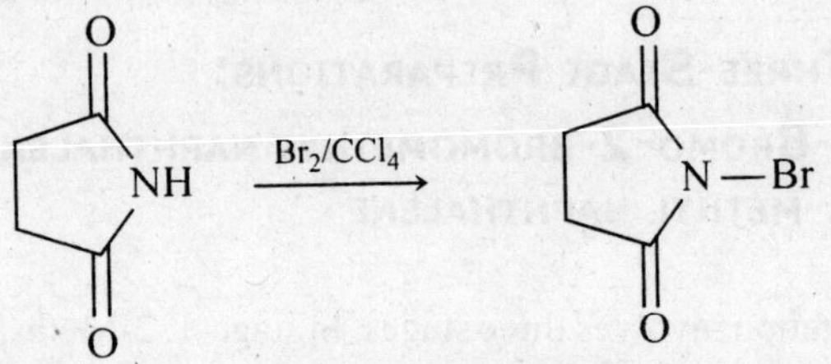

Requirements

Chemicals		Apparatus
Succinimide	16 g	Conical flask
$Bromine/CCl_4$	10 ml	Water bath, stirrer

Procedure

The succinimide is dissolved in sodium hydroxide solution and added to ice. The flask is cooled in an ice bath and bromine is added all at once and the mixture is vigorously stirred till white ppt is obtained. The precipitate is collected by filtration, washed with ice water, till the washings are free of bromine and dired at 40°C.

(Expected yield: 20 g)

Stage 3: Preparation of 1-bromo-2-bromomethyl naphthalene

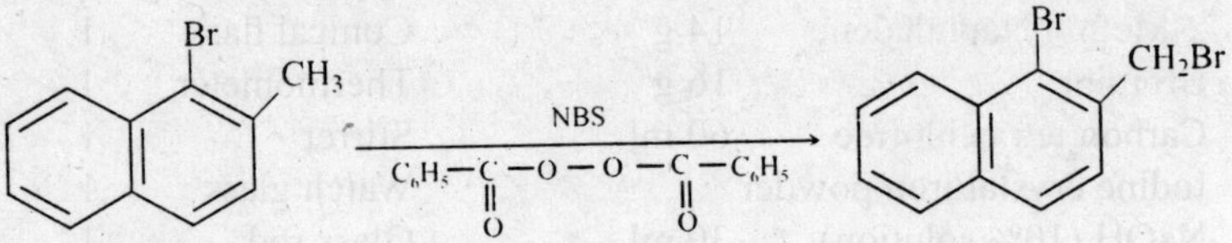

Requirements

Chemicals		Apparatus
1-Bromo-2-methylnaphthalene	20 g	Round Bottom flask
N-Bromosuccinimide	15 g	Funnel
Benzoyl peroxide	0.1 g	Stirrer
Carbon tetrachloride	50 ml	

Procedure

The 1-bromo-2-methylnaphthalene, N-bromosuccinimide, benzoyl peroxide are mixed with 25 ml of carbon tetrachloride in a round bottom flask and the content is refluxed for 3 hours. Another 25 ml of carbon tetrachloride is then added and the mixture is filtered. The solid residue is washed several times with carbon tetrachloride. The combined filtrate is concentrated to about 10 ml and cooled. The product which separates is filtered, washed with water and dried.

(Expected Yield: 20 g; melting point: 104105°C)

(02) SULPHANILAMIDE FROM ACETANILIDE

Principle

Acetanilide on chlorosulphonation, followed by the treatment of ammonia and finally hydrolysis gives sulphanilamide. This reaction takes place in three stages.

(a)Stage-1: Conversion of acetanilide into p-acetamino benzene sulphonyl chloride

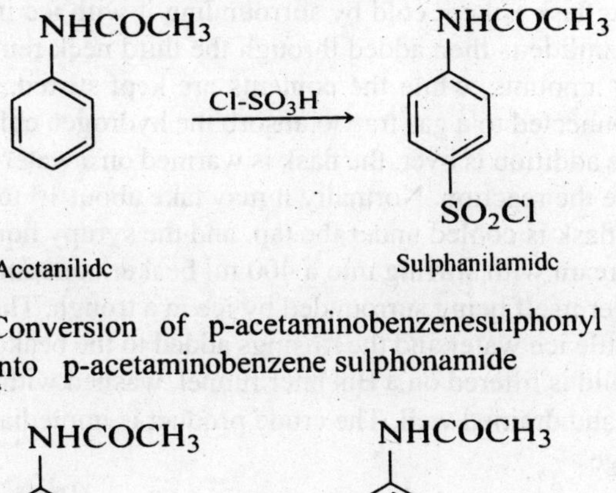

Acctanilidc   Sulphanilamidc

(b)Stage-2: Conversion of p-acetaminobenzenesulphonyl chloride into p-acetaminobenzene sulphonamide

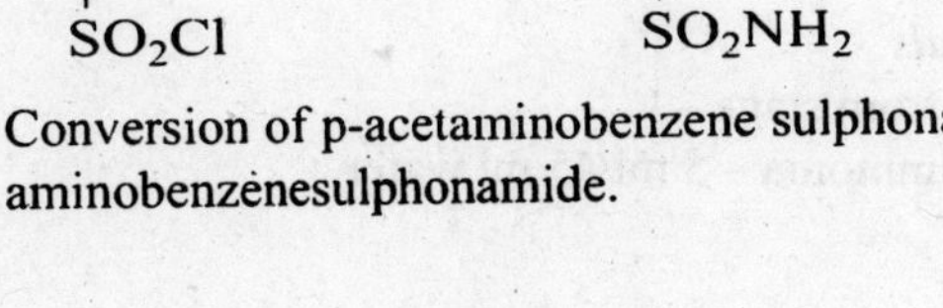

(c) Stage-3: Conversion of p-acetaminobenzene sulphonamide into p-aminobenzenesulphonamide.

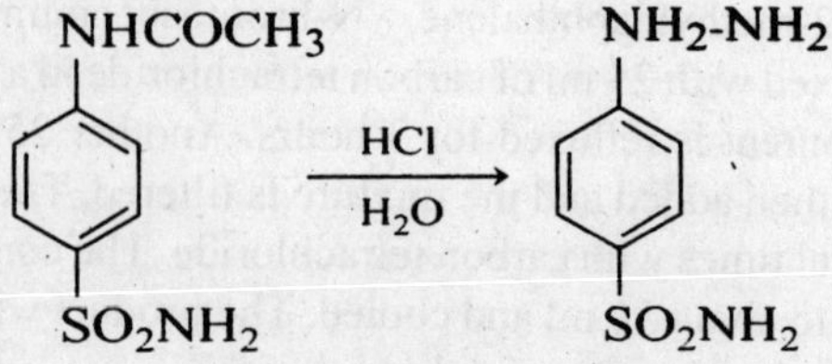

(a) Stage-1: Conversion of acetanilide into p-acetaminobenzenesulphonylchloride

Requirements

Chemicals		***Apparatus***	
Acetanilide	15 g	Three necked flask 250 ml	1
Chlorosulphonic acid	38 ml	Reflux condenser	1
		Water bath, Beaker 400 ml	1

Procedure

The chlorosulphonic acid is introduced into a 250 ml three necked flask fitted with a reflux condenser, and a mechanical stirrer with mercury seal. The flask is kept cold by surrounding it with ice in a trough. The dry acetanilide is then added through the third neck removing the cork in small amounts, while the contents are kept stirred. The condenser top is connected to a gas trap to absorb the hydrogen chloride liberated. After the addition is over, the flask is warmed on a water-bath at 60ºC to complete the reaction. Normally it may take about 15 to 30 minutes.

The flask is cooled under the tap, and the syrupy liquid is poured in a thin stream with stirring into a 400 ml beaker containing crushed ice, the beaker itself being surrounded by ice in a trough. The flask is rinsed with a little ice water and the rinsings added to the beaker. The granular white solid is filtered on a Buchner funnel, washed with ice-cold water, pressed and drained well. The crude product is immediately used in the next stage.

(Expected yield: 20 g)

(b) Stage-2: Conversion into p-acetaminobenzenesulphonamide

Requirements

Chemicals

Product from stage - 1

Liquor ammonia – 5 ml/45 ml water

Procedure

The crude p-acetaminobenzenesulphonyl chloride is transferred to a 500 ml round-bottom flask. Liquor ammonia and 45 ml of water are then added to the flask. The contents of flask are thoroughly mixed and the mixture is heated with occasional swirling to just below the boiling point for about 15 minutes. The sulphonyl chloride is transformed into a paste of sulphonamide. The pasty suspension is cooled in ice and filtered on a Buchner funnel and dried thoroughly.

(Expected yield: 14g)

(c) Stage-3: Conversion of p-acetaminobenzene sulphonamide into p-aminobenzenesulphonamide

Requirements

Chemicals

Conc. HCl-10 ml

Sodium bicarbonate(solid)

Procedure

The product obtained above is transferred into a 250 ml round bottom bolt head flask, 10 ml of concentrated hydrochloric acid and 30 ml of water are added and the mixture is boiled gently under reflux for 30-40 minutes. If the solution, when cooled to room temperature deposits any solid amide, the hydrolysis should be continued further by heating for some more time. The cooled solution is heated to boiling and filtered through a fluted filter paper. The filtrate which is a solution of sulphanilamide hydrochloride is transferred to a beaker and carefully neutralised with solid sodium bicarbonate which is added in small quantities at a time with stirring. The solution is tested with litmus to see that it is neutral. It is then cooled in ice, and the precipitated sulphanilamide filtered and recrystallised from alcohol.

(Expected yield: 10 g; melting point 162°C)

Caution

1. Chlorosulphonic acid is very corrosive and reacts with water violently and hence has to be handled with care.
2. If the sample is impure or coloured it is purified by distillation in an all-glass apparatus. The fraction boiling between 148-150°C is collected, taking care to guard it from moisture.

PART C

SEPARATION OF COMPONENTS OF A BINARY MIXTURE

In Organic Chemistry, separation of components plays an important role, both in qualitative analysis and quantitative estimations. Different techniques are to be adopted for the separation of components from a mixture, depending upon the nature of the components. In these separation experiments, one should aim to obtain the components quantitatively in the pure state. The student must acquire the necessary skill to use the separating funnel.

I. Separation of an Acid and a Hydrocarbon

Principle

The acid components in the mixture is soluble in an excess of aq. NaOH (or aq. sodium carbonate); the neutral compound remains unaffected, if it is insoluble in water. From the sodium salt of the acid solution, the acid is regenerated on acidification with con. HCl.

(01) Separation of Benzoic Acid and Naphthalene

The given solid mixture (about 5-6 g) is taken in a 250 ml beaker. About 20 ml of 10 % aq. NaOH is added, and the mixture is stirred well, and allowed to stand for about 5 minutes. The clear supernatant liquid is decanted into another 250 ml beaker. The addition of aq. NaOH is repeated at least three time, using 20 ml of NaOH at a time, and the process of stirring and decantation is repeated. Practically all the acid component is removed by this process and it remains in solution as the sodium salt. The insoluble hydrocarbon is then washed with water to remove alkali and dried. Naphthalene is thus obtained in an almost pure state. It is further purified by sublimation. Its melting point is determined.

(Correct melting point: 80°C)

The aqueous solution containing the sodium salt of the acid collected in the beaker is treated with con. HCl slowly with constant stirring, till the solution becomes distinctly acidic. The beaker is cooled well under the tap and the precipitated benzoic acid is filtered off at the pump, and washed with cold water. The solid thus obtained is recrystallised from boiling water and the melting point is determined.

(Correct melting point : 121°C.)

(02) Separation of Benzoic Acid and Toluene

The mixture of benzoic acid (about 3 g.) and toluene (about 20 ml) is taken in a clean separating funnel. About 20 ml of 10 % aq. NaOH is added and shaken well. The mixture is allowed to stand for sometime, so that two distinct layers are formed. The lower aqueous layer is collected in a beaker. The addition of aq. NaOH is repeated at least three time using 20 ml of aq. NaOH each time, and the process of shaking is repeated. At each stage, the aqueous layer is collected into the same beaker (for recovering the acid components)

The hydrocarbon layer is then shaken well with 20 ml water, to remove alkali. The aqueous layer is discarded, and the organic layer is collected in boiling test tube. About 2 g. of anhydrous calcium chloride is added to remove any traces of water. When the liquid become clear, it is decanted and distilled. The fraction (toluene) distilling between 108°C and 112°C is collected.

(Correct boiling point : 110°C)

The aqueous solution containing the sodium salt of the acid collected earlier in the beaker is treated with con. HCl slowly with constant stirring, till the solution becomes distinctly acidic. The beaker is then cooled well, and the precipitated benzoic acid is filtered off at the pump, and washed with water. The solid is recrystallised from boiling water and the melting point is determined.

(Correct melting point : 121°C)

II. Separation of an Aromatic Amine and a Hydrocarbon

Principle

The basic component in the mixture (aromatic amine) is soluble in an excess of dil. HCl ; the neutral component is unaffected, if it is insoluble in water. From the aqueous portion, the base is regenerated by treating with aq. NaOH solution.

(03) Separation of Aniline and Toluene

The given mixture (10 ml each of aniline and toluene) is taken in a clean separating funnel, and about 20 ml dil. HCl is added. The mixture is shaken well and allowed to stand for sometime. The lower aqueous layer, containing the amine hydrochloride, is collected in a 250 ml beaker. The addition of dil. HCl is repeated at least three times, using 20 ml of dil. HCl at a time, and the process of shaking is repeated. At each stage, the aqueous layer is collected into the same beaker (for the recovery of the amine).

The hydrocarbon layer is then shaken well with 20 ml water, to remove alkali, and the aqueous layer is discarded. Then, the organic layer is collected in a boiling test tube. About 2 g of anhydrous calcium chloride is added to remove any traces of water. When the liquid becomes clear, it is decanted and distilled. The fraction (toluene) distilling between 108°C and 112°C is collected.

(Correct boiling point: 110°C)

To the aqueous solution of amine hydrochloride collected above, about 20 ml 10 % aq. NaOH is added slowly with constant stirring. The amine separates as a dark liquid. The beaker is colled well, and about 20 ml of diethyl ether is added and shaken well. It is taken in a separating funnel and the lower aqueous layer is removed. The upper ether layer is collected separately. To the aqueous layer another 10 ml of ether is added, shaken well and ether layer is collected from the separating funnel. The aqueous layer is discarded. The combined ether extract is then evaporated carefully by keeping the beaker in a hot water-bath. (*Caution* : Since ether is highly inflammable, avoid bringing ether near a naked flame.) The residual aniline is distilled, and the fraction distilling between 180–185°C is collected.

(Correct boiling point: 184°C.)

III. Separation of a phenol and a Hydrocarbon

Principle

The phenolic component in the mixture is soluble in excess aq. NaOH, the neutral compound remaining unaffected. From the sodium salt of phenolic component, the phenolic compound is regenerated by acidification with con. HCl.

(04) Separation of Phenol and Toluene

The mixture of phenol (about 10 ml) and toluene (about 10 ml) is taken in a clean separating funnel. About 20 ml of 10 % aq. NaOH is added and shaken well. The mixture is allowed to stand for sometime, so that two distinct layers are formed. The lower aqueous layer is collected in a beaker. The addition of aq. NaOH is repeated at least three times, using 20 ml of aq. NaOH each time, and the process of shaking is repeated. At each stage, the aqueous layer is collected into the same breaker (for the recovery of the phenol) (Fraction A).

The hydrocarbon layer is then shaken well with 20 ml water to remove alkali. The lower aqueous layer is discarded, and the upper organic layer is collected in a boiling test tube. About 2 g of anhydrous calcium chloride is added to remove any traces of water. When the liquid becomes clear, it is decanted and distilled. The fraction (toluene) distilling between 108°C and 112°C is collected.

(Correct boiling point: 110°C)

The aqueous layer, containing the sodium salt of phenol, collected earlier (Fraction A) in the beaker is treated with con. HCl, slowly with constant stirring, till the solution becomes acidic. The beaker is cooled well under the tap, and about 20 ml of diethyl ether is added and shaken well. It is then taken in a separating funnel and the aqueous layer is removed. The ether layer is collected in a beaker. The aqueous layer is taken in the separating funnel and about 10 ml of ether is added again to the aqueous layer and shaken well. The ether layer is collected in the same beaker, and the ether is carefully evaporated on a hot water-bath (avoid naked flame). The residual phenol is distilled, and fraction distilling between 177–182°C is collected.

(Correct boiling point: 180°C)

(05) Separation of 2-Naphthol from Toluene

2-Naphthol is a water-insoluble phenolic compound. It dissolves in aq. NaOH giving a soluble sodium salt. From this solution, the naphthol can be precipitated out by the addition of con. HCl.

The separation procedure is exactly the same as that adopted for the separation of benzoic acid from toluene (Separation No.2). 2-Naphthol can be recrystallised from rectified spirit.

(Correct melting point: 122°C)

PART D
ORGANIC ESTIMATIONS

INTRODUCTION

Quantitative estimation of organic compounds plays an important role in the analytical laboratory, *e.g.*, to find out the purity of the chemicals and the extent of adulteration, if any. The quantitative estimations are based on standard chemical reactions and they generally involve volumetric methods. Since most of the organic reactions are slow, instructions regarding mixing of reagents, conducting the reaction, and the time interval, required for titration must be followed strictly.

A knowledge of the chemical reaction involved in the estimation is very essential, in order to understand the calculations involved.

A standard conversion table for glucose estimation is also included.

An error of 2-3 % is generally allowed in organic estimations.

(01) ESTIMATION OF PHENOL

Principle

Phenol on bromination in aqueous solution gives 2,4,6-tribromophenol according to the following equation :—

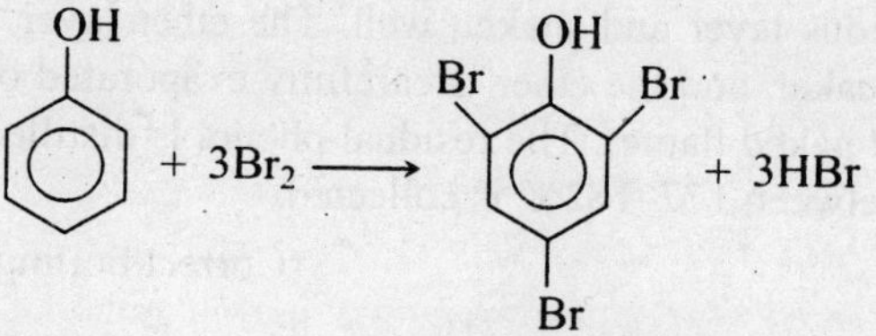

Phenol 2,4,6 - tribromophenol

Instead of using a standard solution of bromine water, a bromate-bromide mixture, which readily liberates bromine in the presence of an acid, is preferred, because its strength does not vary with time.

$$\text{Equivalent weight of phenol} = \frac{\text{Mol.wt}}{6} = \frac{94}{6} = 15.67.$$

The bromate-bromide mixture readily liberates bromine, on acidification, this bromine is used for the bromination of phenol.

$$BrO_3^- + 5Br^- + 6H^+ \longrightarrow 3Br_2 + 3H_2O$$

The aq. soln. of phenol is treated with an excess of the brominating agent, and the unreacted bromine is back-titrated against standard sodium thiosulfate solution using starch as the indicator.

1,000 ml of 1 N thiosulfate = 1,000 ml of 1N Br = 15.67 g of phenol.

1 ml of N/10 thiosulfate = 1 ml of N/10 Br = 0.001567 g. of phenol.

Requirements (Chemials)

AR potassium dichromate—1.2 g. /250 ml (about N/10).

Potassium bromide—3 g
Potassium bromate—0.6 g. } dissolved in 200 ml water (about N/10)

Sodium thiosulfate—6 g. /250 ml water (about N/10).

Potassium iodide—10 g. /100 ml water (about 10%).

Fresh distilled phenol—8 g./1 litre water (about N/2).

(20 ml volume range may be given to each student in a bottle).

Con. HCl—100 ml.

Procedure

Titration 1 : *Standardisation of sodium thiosulfate solution*

A standard solution of N/10 potassium dichromate is prepared thus : about 1.2 g of analar (pure) crystals (equivalent weight = 49), accurately weighed in a chemcial balance, is dissolved in water and the solution is made up to 250 ml in a standard flask.

20 ml of standard $K_2Cr_2O_7$ soln. is pipetted out into a clean conical flask. About 5 ml of con. HCl is added, followed by 10 ml of 10 % aq. KI solution. The liberated iodine is immediately titrated against thiosulfate soln. taken in the burette. When the soln. becomes pale yellow in colour, 1 ml of freshly prepared starch soln. is added, and the titration is continued, adding thio in drops, with constant shaking. The end point is the change of colour from blue to green (due to Cr^{3+}). The titration is repeated to get concordant values. The strength of thiosulfate is calculated.

$$Cr_2O_7^{2-} + 14H^+ + 6I^- \longrightarrow 2Cr^{3+} + 7H_2O + 3I_2$$

$$2Na_2S_2O_3 + I_2 \longrightarrow 2NaI + Na_2S_4O_6$$

Titration 2 : *Standardisation of BrO_3^-/Br^- mixture*

20 ml of the bromate-bromide mixture is pipetted out into a clean 250 ml conical flask. About 10 ml of 10 % aq. KI soln. is added, followed by 5 ml con. HCl. The liberated iodine is titrated against thio taken in the burette, till the solution becomes yellow in colour. About 1 ml of freshly prepared starch soln. is added, and the titration is continued (in drops) with constant shaking. The end point is the disappearance of blue colour (blue to colourless). The titration is repeated to get concordant values.

Titration 3: *Estimation of Phenol*

The given phenol solution in the bottle is quantitatively transferred and made up to 100 ml in a standard flask. 20ml solution is pipetted out into a clean conical flask. The soln. is diluted by adding 50 ml of distilled water. 5 ml con. HCl. is also added. To this a known volume (40 ml) of bromate-bromide mixture is added from the burette, slowly with constant shaking. (The solution should attain a permanent yellow colour, due to the excess of liberated bromine ; otherwise a further known volume of the bromate-bromide is added). Then, about 10 ml of 10 % aq. KI is added. The liberated iodine is titrated against thio using starch as the indicator, which is added near the end point. The end point is the disappearance of blue colour (blue to colourless). A duplicate is also conducted.

Calculation

Calculate the strength of sodium thiosulfate from the standard $K_2Cr_2O_7$ (Titration 1).

Let 20 ml of BrO_3^-/Br^- mixture require A ml of thio (from Titration 2).

Therefore, 40 ml* of BrO_3^-/Br^- should require

$$\frac{A}{20} \times 40^* = 2A \text{ ml. thio.}$$

(* or, actual volume used).

Let (20 ml phenol + 40 ml brominating mix) required B ml thio (Titration 3).

20 ml phenol reacted with (2A – B) ml thio.

$$\text{Strength of phenol solution} = \frac{2A - B \times \text{str.of thio}}{20}$$

$$\left.\begin{array}{l}\text{Amount of phenol in 100 ml}\\ \text{of made up solution}\end{array}\right] = \frac{(2A - B) \times \text{str.of thio}}{20} \times \frac{15.67}{10}$$

$$= \ldots\ldots\ldots \text{ g}$$

Note: Aniline can also be estimated by following the same method as that given above for the estimation of phenol. Aniline reacts with bromine as follows :—

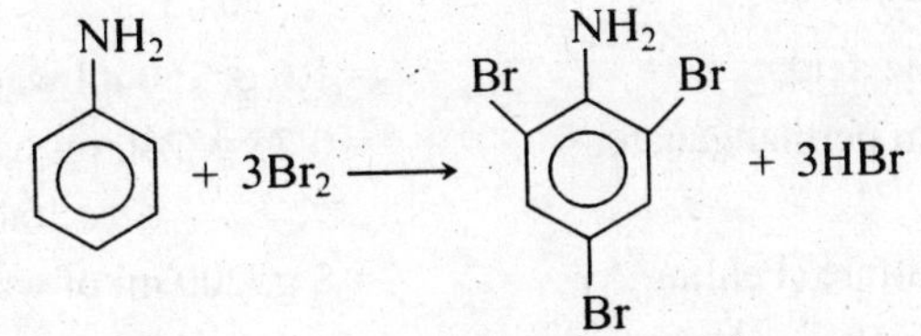

$$\text{Eq. wt. of aniline} = \frac{\text{Molecular weight}}{6} = \frac{93}{6} = 15.5$$

(02) Estimation of Glucose (Bertrand's Method)

Principle

Glucose is a reducing sugar. When glucose is boiled with an excess of alkaline copper hydroxide, it is oxidised to gluconic acid and $Cu(OH)_2$ gets reduced to cuprous oxide.

$$CH_2OH - (CHOH)_4 - CHO + 2\,Cu(OH)_2 \xrightarrow{\text{boiling}}$$

$$CH_2OH - (CHOH)_4 - COOH + Cu_2O \downarrow + H_2O.$$

The precipitated red-coloured Cu_2O is dissolved in a warm acidic solution of ferric alum. Cu_2O reduces ferric sulphate (in ferric alum) to ferrous sulfate. The reduced $FeSO_4$ is titrated against standard $KMnO_4$.

$$Cu_2O + Fe_2(SO_4)_3 + H_2SO_4 \rightarrow 2CuSO_4 + 2\,FeSO_4 + H_2O$$

$$10FeSO_4 + 8H_2SO_4 + 2KMnO_4 \rightarrow K_2SO_4 + 2\,MnSO_4 + 5\,Fe_2(SO_4)_3 + 8\,H_2O.$$

2 Cu ≡ 2 $FeSO_4$	≡ 2 litres of 1 N $KMnO_4$.
1 atom of copper (63.6. g)	≡ 1 litre of 1 N $KMnO_4$.
1000 ml of 1 N $KMnO_4$	≡ 63.6 g. of copper.
1 ml of N/10 $KMnO_4$	≡ 0.00636 g. of copper.
	≡ 6.36 mg. of coper.

From the amount of copper, the quivalent amount of sugar is calculated from the given conversion table.

Chemicals required

AR Oxalic acid	= 1.6 g./250 ml water.
Potassium permanganate	= 0.75 g./250 ml water (about N/10).
Cupric sulfate (Fehling A)	= 8 g./200 ml of water
Rochelle salt — 40 g, Caustic soda — 30 g (Fehling B)	= dissolved in 200 ml water.
Ferric alum–24 g., Con. H_2SO_4—20 ml	= dissolved in 200 ml
Dil. H_2SO_4 — 100 ml AR Glucose	= 13g./1 litre of water.

(20 ml volume range may be given to each student in a bottle.)

Procedure

Titration 1: *Standardisation of potassium permanganate solution*

A standard soln. of N/10 oxalic acid is prepared by weighing about 1.6 g. of analar acid crystals (equivalent weight : 63) accurately in a chemical balance and dissolving in water. The soln. is then made upto 250 ml in a standard flask.

20 ml of standard oxalic acid soln. is pipetted out into a clean 250 ml conical flask. 20 ml dilute sulphuric acid is added 60°C. (bearable warmth). The hot soln. is then titrated against $KMnO_4$ soln. taken in the burette. The end point is the appearance of pale permanent pink colour. The titration is repeated to get concordant values. The strength of $KMnO_4$ is calculated.

$$2MnO_4^- + 16H^+ + 5C_2O_4^{2-} \longrightarrow 2Mn^{2+} + 10CO_2 + 8H_2O$$

Titration 2 : *Estimation of Glucose*

The given glucose solution in the bottle is quantitatively transferred and made up to 100 ml in a standard flask. 20 ml is pipetted out into a clean 250 ml conical flask. About 20 ml of $CuSO_4$ solution (Fehling solution-A) is added, followed by the addition of 20 ml of tartrate solution (Fehling solution-B). The solution is then heated to boiling, and maintained at this temperature for 2 to 3 minutes. The red cuprous oxide precipitate formed is allowed to settle. The clear supernatant liquid is carefully decanted.

Amount of Copper precipitated by different amounts of Glucose

Glucose in mg.	*Copper in mg.*	*Glucose in mg.*	*Copper in mg.*	*Gluose in mg.*	*Copper in mg.*
10	20.4	41	79.3	71	31.4
11	22.4	42	81.1	72	33.1
12	24.3	43	82.9	72	34.7
13	26.3	44	84.7	74	36.3
14	28.3	45	86.4	75	37.9
15	30.2	46	88.2	76	39.6
16	32.3	47	90.0	77	41.2
17	34.2	48	91.8	78	42.8
18	36.2	49	93.6	79	44.5
19	38.1	50	95.4	80	46.1
20	41.1	51	97.1	81	47.7
21	42.0	52	98.9	82	49.3
22	43.9	53	100.6	83	50.9
23	45.8	54	102.3	84	52.5
24	47.7	55	104.1	85	54.0
25	49.6	56	105.8	86	55.6
26	51.5	57	107.6	87	157.2
27	53.4	58	109.3	88	158.8
28	55.3	59	111.1	89	160.4
29	57.2	60	112.8	90	162.0
30	59.1	61	114.5	91	163.6
31	60.9	62	116.2	92	165.2
32	62.8	63	117.9	93	166.7
33	64.6	64	119.6	94	168.3
34	66.5	65	121.3	95	169.9
35	68.3	66	123.0	96	171.5
36	70.1	67	124.7	97	173.1
37	72.0	68	126.4	98	174.6
38	73.8	69	128.1	99	176.2
39	75.7	70	129.8	100	177.8
40	77.5				

The precipitate is washed with water to remove the excess of $CuSO_4$ solution (It is advisable to filter the precipitate through G-4 sintered glass crucible.) To the precipitate, acidified ferric alum is added slowly with shaking, till the red precipitate completely dissolves. About 20 ml of dil. H_2SO_4 is added and the soln. is titrated against $KMnO_4$ taken in the burette. The end point is the appearance of pale permanent pink colour. A duplicate is also conducted.

Calculation

Caculate the strength of $KMnO_4$ from the standard oxalic acid (Titration 1).

Volume of sugar solution taken = 20 ml

Volume of $KMnO_4$ required to react with the reduced $FeSO_4$ = A ml

1 ml of N/10 $KMnO_4$ = 6.36 mg of copper

$$\text{A ml of standard } KMnO_4 = \frac{6.36 \times A \times \text{str.of } KMnO_4}{0.1}$$

= B mg of copper.

From the conversion table, the glucose equivalent is noted, corresponding to copper equivalent.

B mg. of Cu = Z mg of sugar

20 ml glucose solution contains Z mg of sugar.

$$\left.\begin{array}{l}\text{Amount of glucose present in}\\ \text{100 ml of the made-up solution}\end{array}\right\} \frac{100}{20} Z = 5\,Z \text{ mg}$$

(03) Iodine Value on an Oil (Hanus Method)

Principle

Iodine value is defined as the number of parts by weight of iodine reacting with 100 parts by weight of an oil or fat. The drying power of an oil is generally proportional to its iodine value. Coconut oil (a non-drying oil) has low iodine value of 6–10 ; linseed oil (a drying oil) has high iodine of 77–88 value. Idoine value indicates the degree of unsaturation of the fatty acids present in the oil or fat. The determination of the iodine value is of great importance in characterising an oil, and al.o in finding the proportion of an adulterant in a sample of the oil.

In Hanus method of determination of the iodine value, a known weight of oil is dissolved in CCl_4, and treated with a known excess volume of iodine monobromide solution. The unused IBr is backtitrated against standard sodium thiosulfate solution.

Iodine Values of some common oils are given below :

(1) Coconut oil = 6–10.	(5) Cotton seed oil = 103–111.
(2) Olive oil = 79–88.	(6) Gingelly oil = 103–117.
(3) Castor oil = 84.	(7) Mustard oil = 105–110.
(4) Groundnut oil = 83–100.	(8) Linseed oil = 175–202.

Chemicals required

AR potassium dichromate = 1.2 g / 250 ml water.
Sodium thiosulfate = 6 g / 250 ml water (about N/10)
Potassium iodide = 10 g / 100 ml water.
Starch solution (freshly prepared) = 10 ml
IBr - Iodine (13.2 g./ 1 litre HOAc) + 3 ml liquid bromine.
(100 ml required for each student.)
Carbon tetrachloride = 20 ml.

Procedure

Titration 1 : *Standardisation of Thio.*

A standard soln. of N/10 $K_2Cr_2O_7$ is prepared by weighing about 1.2 g of analar crystals (eq. wt. = 49) accurately in a chemical balance, dissolving in water and making up to 250 ml in a standard flask.

20 ml of standard $K_2Cr_2O_7$ soln. is pipetted out into a clean 250 ml conical flask. About 5 ml con. HCl is added, followed by 10 ml of 10 % aq. KI soln. The liberated iodine is immediately titrated against thio taken in the burette. When the solution acquires a straw-yellow colour, 1 ml of freshly prepared starch is added and the titration is continued (in drops) with constant shaking. The end point is the change in colour from blue to green (due to Cr^{3+}). The titration is repeated to get concordant values. The strength of thio is calculated.

$$Cr_2O_7^{2-} + 14H^+ + 6I^- \longrightarrow 2Cr^{3+} + 7H_2O + 3I_2$$

$$2S_2O_3^{2-} + I_2 \longrightarrow 2I^- + S_4O_6^{2-}$$

Titration 2 : *Determination of Iodine value.*

About 0.5 to 1 g. of the given oil is weighed accurately in a chemical balance and dissolved in 10 ml of CCl_4 in a stoppered 500 ml bottle.

25 ml of iodine monobromide solution is buretted out into the bottle and the time is noted. The resulting mixture, if turbied is cleared by adding a small additional known volume of CCl_4. The stopper is moistened with a few drops of aq. KI solution and then inserted into the bottle. The bottle is kept aside for about 40–60 minutes, with occasional shaking. Then, the reaction mixture is diluted with 200 ml of water, followed by the addition of 20 ml of 10 % aq. KI soln. The mixture is titrated against thio taken in the burette, using starch as the indicator, added near the end point (disappearance of blue colour). A duplicate is also conducted.

A blank titration is carried out without the oil, using exactly the same quantity of CCl_4 (10 ml plus any additional amount used) and 25 ml of iodine monobromide soln., using starch as indicator.

Calculation

Calculate the strength of thio.

Let A ml be the volume of $Na_2S_2O_3$ required for the excess IBr soln.

Let B ml be the blank titre value.

Iodine value = $(B - A) \times \frac{127}{1000} \times \frac{100}{w} \times$ str. of thio, where w is the weight of oil taken, and 127 is the atomic weight of iodine.

(04) SAPONIFICATION VALUE OF AN OIL

Principle

The saponification value is defined as the number of milligrams of KOH required for the hydrolysis of 1 g. of an oil or a fat. To determine the saponification value, a weighed quantity of the given oil is reflexed with a known volume of alcoholic potassium hydroxide soln. The unused alkali is then back-titrated against standard acid.

Saponification values of some common oils are given below

(1)	Mustard oil = 174.	(5)	Linseed oil = 188–195.
(2)	Castor oil = 175–183.	(6)	Gingelly oil = 188–193.
(3)	Olive oil = 185–196.	(7)	Cotton seed oil = 194–196.
(4)	Groundnut oil = 186–194.	(8)	Coconut oil = 253–262.

Chemicals required

AR sodium carbonate crystals	=	7 g. / 250 ml water.
Alcoholic potash (KOH)	=	6 g dissolved in 200 ml rectified spirit (about 0.5 N)

Phenolphthalein

HCl (about N/2) = 10 ml con. HCl in 200 ml water.

Procedure

Titration 1: *Standardisation of HCl.*

About 7 g. of analar sodium carbonate crystals (eq. wt = 53) are weighed accurately in a chemical balance. dissolved in water, and the solution in made upto 250 ml in standard flask. 20 ml is pipetted out into a clean concical flask. About 2 drops methyl orange indicator are added and titrated against hydrochloric acid taken in the burette (till the light pink colour appears). The titration is repeated to get concordant values. The strength of HCl soln. is calculated.

Titration 2: *Determination of saponification value of an oil.*

About 1–2 g of the given oil is weighed accurately in a chemical balance and transferred into 250 ml round-bottom flask. 25 ml of alcoholic potash (about N/2) is added from the burette slowly into the flask. It is then fitted with an air condenser, and heated on a water-bath for about 30 minutes. A blank is run simultaneously with the same quantity of alcoholic potash, but without the oil. Both the flasks are cooled, and titrated against standard HCl soln. using about 1 ml of phenolphthalein indicator. The end point is the disappearance of pink colour.

Calculation

Calculate the strength of HCl soln. (Titration 1.)

Let *w* g. be the weight of oil taken.

Let A ml be the volume of HCl required for excess alcoholic potash.

Let B ml be the blank titre value.

1000 ml of 1 N HCl	=	56.1 g. of KOH.
1 ml 1 N HCl	=	56.1 mg. of KOH.

$$\text{Saponification value} = \frac{56.1 \times (B - A) \times \text{str. of HCl}}{w}$$

(05) ESTIMATION OF ACETONE

Principle

Acetone (or any ketone containing CH_3–CO group) forms iodoform, when shaken well with a mixture of iodine and aq. KOH soln. according to the following equations

$$I_2 + 2\,KOH \rightarrow KI + KOI + H_2O$$

$$CH_3COCH_3 + 3\,KOI \rightarrow CI_3COCH_3 + 3\,KOH$$

$$CI_3COCH_3 + KOH \rightarrow CHI_3 + CH_3COOK.$$

3 molecules of iodine = 1 molecule of acetone.
3 molecules of iodine = 58 g of acetone.
6 litres of 1 N iodine = 58 g of acetone.
1 ml of 1 N iodine = 0.0097 g of acetone.

The excess of iodine (present as KI and KOI) is decomposed in the presence of an acid, according to the following equation

$$KI + KOI + H_2SO_4 \rightarrow K_2SO_4 + I_2 + H_2O$$

The liberated iodine is titrated against standard thio using starch as indicator.

$$I_2 + 2Na_2S_2O_3 \rightarrow 2\,NaI + Na_2S_4O_6 \text{ (sod. tetrathionate)}$$

Chemical required

AR potassium dichromate	=	1.2 g./250 ml water.
Sodium thiosulfate	=	6 g. / 250 ml of water (about N/10).
Potassium iodide (10 %)	=	10 g. / 100 ml water.
Starch solution (1 %)	=	10 ml (freshly prepared).
Con. HCl	=	20 ml
Potassium hydroxide	=	11 g. / 200 ml water (about 1 N).
Aq. iodine	=	2.5 g. / 200 ml of water (about N/10).
Acetone	=	5 g. / 1 lit. water (about N/2)

(20 ml volume range may be given to each student in a bottle.)

Procedure

Titration 1 : *Standardisation of thio.*

A standard soln. of N/10 $K_2Cr_2O_7$ is prepared by weighing about 1.2 g. of analar crystals (eq. wt. = 49) accurately in a chemical balance, dissolving in water, and making up to 250 ml in a standard flash.

20 ml of standard $K_2Cr_2O_7$ soln. is pipetted out into a 250 ml clean conical flask. About 5 ml con. HCl is added, followed by 10 ml of 10% aq. KI. The liberated iodine is immediately titrated against thio taken in the burette. When the soln. acquires straw yellow colour, 1 ml of freshly prepared starch is added, and the titration is continued (in drops) with constant shaking. The end point is the change in colour from blue to green (due to Cr^{3+}). The titration is repeated to get concordant values. The strength of thio is calculated.

$$Cr_2O_7^{2-} + 14H^+ + 6I^- \rightarrow 2Cr^{3+} + 7H_2O + 3I_2$$

$$2S_2O_3^{2-} + I_2 \rightarrow 2I^- + S_4O_6^{2-}$$

Titration 2 : *Standardisation of Iodine soln.*

20 ml of iodine soln. is pipetted out into a clean 250 ml conical flask. About 1 ml starch is added. The soln. is titrated against thio taken in the burette. The end point is the disappearance of blue colour (blue to colourless). The titration is repeated to get concordant values.

Titration 3 : *Estimation of acetone.*

The given acetone solution in the bottle is quantitatively transferred and made up to 100 ml in a standard flask. 20 ml of the made-up soln. is pipetted out into a clean 500 ml stoppered bottle. 50 ml of 1 N aq. KOH is added to the bottle. A known volume of aq. iodine (100 ml of N/10) is added from the burette, slowly, into the bottle, and the mixture is shaken well for 15 minutes. It is then acidified with about 20 ml of dil. H_2SO_4 and the liberated iodine is titrated against thiosulfate solution taken in the burette, using starch as indicator. The end point is the disappearance of blue colour.

A blank titration is carried out without acetone, using the same volume of KOH and iodine against sodium thiosulfate, using starch as indicator.

Calculation

Calculate the strengths of thio and iodine solutions.
(Titrations 1 and 2)
Let the volume of thio required by

(acetone + I_2) solution = A ml

Blank titre value (100 ml iodine + KOH) = B ml.
B ml of thio is required for 100 ml iodine.

A ml of thio is required for $\frac{100}{B} \times A$ = x ml of iodine.

Unreacted iodine = x ml.

Reacted iodine = $(100 - x)$ ml.

1 ml of N iodine = 0.0097 g. of acetone.

$(100 - x)$ ml of iodine = $0.0097 \times (100 - x)$ × (str. of iodine) g. of acetone.

100 ml of made-up soln contains = $0.0097 \times (100 - x) \times (\text{str. of } I_2) \times \frac{100}{20}$

= ...... g. of acetone.

[**Note:** Ethyl methyl ketone may also be estimated using the same method]

(06) Estimation of Ascorbic Acid (Vitamin C)

Principle

Ascorbic acid can be oxidised to dehydro-ascorbic acid by iodine, according to the equation:

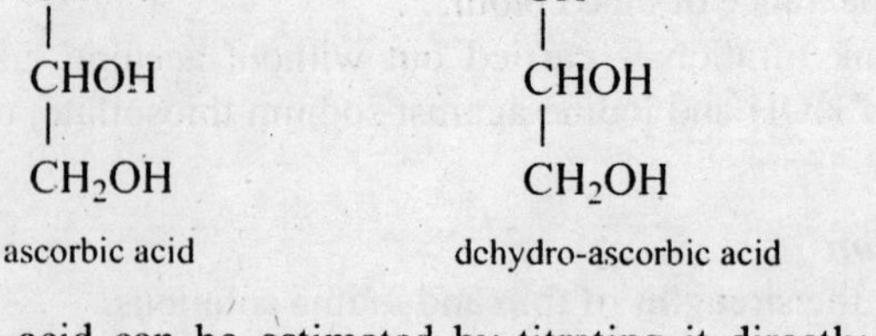

ascorbic acid dehydro-ascorbic acid

Ascorbic acid can be estimated by titrating it directly against a ındard soln. of iodine, using starch as indicator.

q. wt. of ascorbic acid = 88.)

Chemicals required

AR potassium dichromate = 1.2 g / 250 ml water (about N/10).
Sodium thiosulfate = 5 g / 200 ml of water (about N/10).
Iodine solution = 2.6 g / 200 ml of water (about N/10).
Potassium iodide (10 %) = 10 g / 100 ml of water.
Starch solution (1 %) = 10 ml (freshly prepared).
Pure ascorbic acid = 44 g / 1 litre of water (about N/2).
(20 ml volume range may be given to each student in a bottle.)

Procedure

Titration 1: *Standardisation of Thio.*

A standard soln. of N/10 $K_2Cr_2O_7$ is prepared by weighing about 1.2 g. of analar crystals (eq. wt. = 49) accurately in a chemical balance, dissolving in water and making up to 250 ml in a standard flask.

20 ml of standard $K_2Cr_2O_7$ soln. is pipetted out into a clean conical flask. About 5 ml con. HCl is added, followed by 10 ml of 10 % aq. KI. The liberated iodine is immediately titrated against thio taken in the burette. When the soln. acquires a straw yellow colour, 1 ml of freshly prepared starch added, and the titration is continued (in drops) with constant shaking. The end point is the colour change from blue to green (due to Cr^{3+}). The titration is repeated to get concordant values. The strength of thio is calculated.

$$Cr_2O_7^{2-} + 14H^+ + 6I^- \longrightarrow 2Cr^{3+} + 7H_2O + 3I_2$$

$$2S_2O_3^{2-} + I_2 \longrightarrow 2I^- + S_4O_6^{2-}$$

Titration 2: *Standardisation of iodine solution.*

20 ml of iodine soln. is pipetted out into a clean 250 ml conical flask. About 1 ml starch is added. The soln. is titrated against thio taken in the burette. The end point is the disappearance of blue colour (blue to colourless). The titration is repeated to get concordant values.

Titration 3: *Estimation of ascorbic acid.*

The given ascorbic acid solution in the bottle is quantitatively transferred into a 100 ml std. flask and made-up to the mark with distilled water. 20 ml of this soln. is pipetted out into a 250 ml clean conical flask, 1 ml of freshly prepared starch is added, and the resulting soln. is titrated against iodine soln. taken in the burette. The end point is the appearance of blue colour. The titration is repeated to get concordant values.

Calculation

From the strength of iodine soln., the strength of ascorbic acid, and hence the amount present in the given solution is calculated.

(07) Estimation of Sulphur (Carius' Method)

Principle

A known weight of the substance is heated with fuming nitric acid in a sealed tube, the organic material being oxidized to carbon dioxide and water, and sulphur to sulphuric acid. The latter is subsequently washed out of the tube, precipitated as barium sulphate and estimated as barium sulphate. The method is general for all organic compounds.

Procedure

A known mass(about 1g) of an organic compound is taken(ofcourse the amount depends on the % of S such that the weight of $BaSO_4$ should be about 0.2-0.3 g) in a hard glass test tube and mixed with fuming nitric acid. The test tube is then sealed on the top and placed in a heater or in a oil bath and maintained a temperature of 260°C for 2 to 3 hours. The Carius tube is then cooled. Before opening the cold Carius tube, pay particular attention to the initial gentle warming of the capillary end of the tube to ensure that the liquid which has condensed in the capillary is driven back well into the wide portion of the tube, otherwise it will be lost with the escaping gases. The tube is then cut open and the contents are transferred into a beaker. The tube is then washed well with water several times and all the washings are collected in the same container. The solution is then diluted to 200 ml and heated. To the hot solution, about 30 ml of 5% barium chloride solution is added very slowly with constant stirring. (The precipitation of the barium sulphate must be performed with care, otherwise high results are obtained owing to occlusion of barium chloride in the barium sulphate). The solution is then digested on a water bath for 30 minutes. The barium sulphate precipitated is filtered off at the pump using weighed sintered glass crucible, washed with water, dried in hot air oven at 130°C and weighed. From the mass of barium sulphate precipitate, the % of Sulphur in the organic compound is calculated gravimetrically.

Calculation

Let the mass of barium sulphate = x g

233 g of barium sulphate contains 32 g of Sulphur

x g of barium sulphate contains $\frac{32}{233} \times x$ g of Sulphur

w g of organic substance contains $\frac{32}{233} \times x$ g of Sulphur

$\therefore$ 100 g of organic substance contains $\frac{32}{233} \times x \times \frac{100}{w}$ %

Note: Instead of con. HNO_3 as an oxidizing agent, alkaline $KMnO_4$ can also be used to oxidise Sulphur in the organic compound into sulphuric acid quantitatively. For this about 2 g of solid $KMnO_4$ and 1g KOH can be used.

(08) Estimation of Amino Group

Determination of the percentage of amino groups in an organic amino compound by acetylation method

Principle

Organic amino compounds when treated with a known amount of acetic anhydride or acetyl chloride forms acetyl derivative.

$$R-(NH_2)_n + n[(CH_3CO)_2O] \rightarrow R\text{-}(NH-CO-CH_3)_n + nCH_3COOH$$

The excess of acetic anhydride or acetyl chloride is then hydrolysed with water and the total free acetic acid is treated with a standard sodium hydroxide solution using phenolphthalein as an indicator.

A blank experiment is performed (without using amino compound) with similar amount of acetylating agent.

The difference in the amount of sodium hydroxide used in the two experiments is equivalent to the acetic acid used in acetylation.

Chemicals required

Organic amino compound	1g
Acetic anhydride	10 ml
Pure anhydrous pyridine	40 ml
Sodium hydroxide (N/5)	250 ml
Phenolphthalein	

Procedure

Standardisation of NaOH: N/5 Sodium hydroxide solution is standardized using a standard oxalic acid solution.

Preparation of acetylation mixture: Acetic anhydride and pyridine are mixed in a dry conical flask. The burette is filled with this solution, fitted with an anhydrous calcium chloride tube at the top.

Acetylation of amino compound: About 0.5g of the organic compound is accurately weighed and transferred into a 100 ml conical flask fitted with a reflux water condenser. The acetylating agent is then slowly added to the conical flask and the mixture is then heated on a boiling water bath for 30 minutes. About 25 ml of water is then added through the condenser tube. The mixture is shaken well for 5-10 min. The flask is cooled to room temperature.

Titration of free acetic acid with sodium hydroxide: The burette is filled with sodium hydroxide solution. 20 ml of the reaction mixture is pipetted out into a clean conical flask. Two to three drops of phenolphthalein is added as an indicator. The content is titrated with sodium hydroxide solution till pink colour just persists on shaking.

Blank titration: A blank titration is also conducted following the above procedure without the amino compound.

Calculation

The mass of amino compound = w g

Let the blank titre value be X ml

20 ml of hydrolysed acetylating agent requires X ml of N/5 NaOH

Let the test titre value be Y ml

20 ml of amino compound + acetylating agent requires Y ml of N/5 NaOH

∴ Volume of N/5 NaOH required for 20 ml of amino compound = (Y – X) ml

1000 ml of 1N NaOH = 1 mole of NaOH

= 1 mole of CH_3COOH

= 1 amino group

$$(Y - X) \text{ ml of N/5 NaOH} = \frac{1}{1000} \times (Y - X) \times \frac{N}{5}$$

$$\text{w g of amino compound contains } \frac{1}{1000} \times (Y - X) \times \frac{N}{5} \text{ amino groups}$$

100 g of amino compound contains $\frac{1}{1000} \times (Y - X) \times \frac{N}{5} \times \frac{100}{w}$ amino groups

(09) Estimation of Nitro Group

Principle

The estimation is based on the reduction of the nitro compounds to amino compounds when treated with an excess of a reducing agent. The excess of the reducing agent is then estimated. Stannous chloride is used ordinarily as the reducing agent.

$$R - NO_2 + 3SnCl_2 + 6HCl \rightarrow RNH_2 + 3SnCl_4 + 2H_2O$$

Hence $NO_2 = 3\ SnCl_2 = 6$ litres of 1N solution,
since $SnCl_2 \equiv I_2 \equiv 2$ litres of 1N solution.

Preparation of stannous chloride solution: 20 g of pure tin powder or granules are introduced into a 250 ml round-bottom flask fitted with an air condenser. 200 ml of concentrated hydrochloric acid are added and the flask is heated till the reaction starts. When all the tin has dissolved the clear solution is decanted from undissolved matter and made up to a little with distilled water.

Procedure

About 0.5g of the non-volatile nitro compound is introduced into a bolt-head flask of 250 ml capacity, 30 ml of alcohol are added and the flask gently warmed to dissolve the compound. The flask is fitted with a two-holed cork, carrying a leading-in tube for passage for carbon dioxide and a reflux condenser. The air in the flask is displaced by carbon dioxide and a known volume of stannous chloride (50 to 75 ml) is added after removing the cork. The cork with the condenser and the leading-in tube is quickly replaced and the flask is heated over a wire gauze for about two hours while a stream of carbon dioxide is bubbling through.

Meanwhile, the strength of the original stannous chloride is determined by titration with standard iodine using starch as indicator in a flask through which a current of carbon dioxide is passing to prevent atmospheric oxidation.

The excess of stannous chloride is determined by titration with standard iodine in the same manner.

$$xNO_2 + 3SnCl_2 + 6HCl \rightarrow xNH_2 + 3SnCl_4 + 2H_2O$$

no. of $NO_2 \equiv 3\ SnCl_2 \equiv 6$ litres of normal solution,
since $SnCl_2 \equiv I_2 \equiv 2$ litres of normal solution.

Let the volume of $SnCl_2$ taken = V ml of exactly normal iodine.
Let excess of $SnCl_2$ = V_1 ml of exactly normal iodine
$SnCl_2$ used up = $(V - V_1)$ ml of 1N iodine solution

$$\% NO_2 = \frac{(V - V_1) \times 46 \times 100}{6000 \times m}$$

where m = weight of nitro compound.

From the percentage of nitro group the weight of the substance which contains one nitro group (46 g of it) can be calculated.

Then the number of nitro groups present

$$= \frac{\text{Molecular weight}}{\text{weight of nitro compound}}$$

The following compounds may be used for estimating the nitro group:

1. m-Dinitrobenzene 2. p-Nitrotoluene 3. p-Nitrochlorobenzene
4. Nitrobenzene 5. o-Nitrotoluene

(10) Estimation of Acetyl Groups (Perkin's Method)

Principle

The acetyl derivative in an organic compound is hydrolysed in the presence of alcohol with concentrated sulphuric acid and the ethyl acetate formed is absorbed in standard alcoholic potash.

$$ROCOCH_3 + C_2H_5OH \rightarrow ROH + CH_3COOC_2H_5$$

1 equivalent of a base ≡ 1 equivalent of acetyl group
1 litre of N NaOH ≡ 43 g of acetyl group
1 ml of N NaOH ≡ 0.043 g of acetyl group

Procedure

A distilling flask (250 ml) with its side tube bent is fitted to a long water-cooled condenser. About 0.5 g of the acetyl derivative is taken in an ignition tube and the substance transferred with the tube slightly inclined so that it falls directly on to the bottom of the flask. The difference in weight of the tube gives the weight of the substance transferred.

Any particles adhering to the sides are washed down by 5 ml of strong sulphuric acid delivered from a pipette, while the flask is gently swirled and cooled under the tap.

A piece of porous porcelain is added and an adapter is attached to the condenser tube. The flask is fitted with one holed rubber stopper carrying a tap funnel. 20 ml of alcohol are placed in the funnel and 20 ml of N/2 alcoholic potash in the 200 ml round bottom flask surrounding the adapter. The flask is heated on a wire gauze with a Bunsen burner carrying a nose tap, and the alcohol is added at such a rate, that it equals the rate of distillation. The distillation is continued till about half the liquid has distilled over. There should be no carbonization taking place. The receiver is attached to a vertical water cooled condenser and boiled on the water bath for about half an hour, after which, the inner tube of the condenser is washed with a little water and the excess of alkali titrated against standard acid using phenolphthalein as indicator.

A blank titration is also carried out for the standardization of alkali using a standard acid.

Calculation

1 litre of N NaOH ≡ 43 g of acetyl group
1 ml of N NaOH ≡ 0.043 g acetyl group

Let 20 ml of N/2 alkali be = V_1 ml of 1N HCl.
Let the titre value of the reaction be V_2 ml
$(V_1 - V_2)$ = V ml 1N acid in alkali = w g of the solution

$$\% \text{ of acetyl group in the compound} = \frac{V \times 0.043}{w} \times 100$$

No. of acetyl groups ≡ M/x where M = molecular weight and x = wt. of compound containing 1 acetyl group.

Note: 1 The method is not applicable to acetyl derivatives of amines.

2. The following compounds may be used for estimating the acetyl group:
1. β–Naphthyl acetate 2. Hydroquinone diacetate
3. Pyrogallol triacetate 4. Acetyl vanillin

(11) Estimation of Methoxyl Groups (Perkin's Modified Zeisel's Method)

Principle

The organic compound containing methoxyl group is heated with concentrated hydroiodic acid so produced is carried over by a stream of carbon dioxide and absorbed in a solution of alcoholic silver nitrate. The silver iodide precipitated is then weighed.

$$R(OCH_3)_n + nHI \rightarrow R(OH)_n + nCH_3I$$

1 mole of $-OCH_3$ group ≡ 1 mole of CH_3I ≡ 1 mole of AgI

235 g of AgI precipitated ≡ 31 g of methoxyl group

By knowing the mass of AgI precipitated, the mass of methoxyl group and hence the number of methoxyl group can be calculated.

Preparation of solutions

Hydroiodic acid: The required density of the HI acid is 1.7 corresponding to the constant boiling mixture, b.p. 127°C. For every 25 g of the acid 4 ml of water are to be added. Red phosphorous is likely to give off volatile impurities and therefore is boiled with 10% caustic potash till free from phosphoretted hydrogen, then filtered, washed free from alkali and preserved in distilled water.

Alcoholic silver nitrate: 2 g of silver nitrate are dissolved in 5 ml of water and then diluted with 45 ml of absolute alcohol. The solution is preserved in a well stoppered bottle in the dark. It should be filtered and acidified with one drop of nitric acid immediately before use. For each estimation, 20 ml of the solution are taken.

Procedure

0.5 g of the organic substance is weighed accurately into a small tube. About 15 ml of the hydriodic acid are introduced into the Zeisel's flask.

and the tube containing the substance is gently slid into the flask. A few pieces of porous porcelain are added.

A very slow stream of carbon dioxide purified by passing through water and dried by concentrated sulphuric acid is passed at the rate of two bubbles per second, the leading in tube being just over the surface of liquid instead of actually dipping into it. The flask is heated in a glycerine or metal bath gradually to 130°C. The methyl iodide passing over, is freed from HI and iodine by a wash bottle containing red phosphorous suspended in water, and precipitates silver iodide in the first absorption flask. When the upper portion of the liquid in the first flask is clear after the silver iodide has settled, the temperature of the bath is raised, so as to boil very gently the hydroiodic acid. After an hour, the absorption flasks are disconnected and the gas led into a test tube containing some of the silver nitrate. If no precipitate is seen in about five minutes the reaction is complete. The contents of the two wash bottles are diluted with a little water, kept for a few minutes to ensure complete decomposition of the last traces of methyl iodide, and finally poured into 50 ml of boiling water acidified with nitric acid and boiled to drive off the alcohol. The precipitate is filtered in a glass crucible with sintered septum (1 G 4) and dried at 120°C.

Calculation

1 mole of $-OCH_3$ group ≡ 1 mole of CH_3I ≡ 1 mole of AgI

235 g of AgI precipitated ≡ 31 g of methoxyl group

Let weight of the organic substance taken = W g

Let the weight of AgI formed = W_1 g

235 g of AgI precipitated ≡ 31 $W_1 \times \frac{31}{235}$ g of methoxyl group

$\therefore W_1$ g of AgI precipitated ≡ $W_1 \times \frac{31}{235}$ g of methoxyl group

W g of organic substance contains $\left(W_1 \times \frac{31}{235}\right)$ g of methoxyl group

$\therefore$ 100 g of organic substance contains $\left(\frac{W_1}{W} \times \frac{31}{235} \times 100\right)$ g of methoxyl group

$$\% \text{ Methoxyl group} = \frac{W_1}{W} \times \frac{31}{235} \times 100$$

(12) ESTIMATION OF ALDEHYDES

(Estimation of formaldehyde in commercial formalin)

Principle

Formaldehyde is oxidsied into formic acid on treatment with iodine in alkaline medium. From the number of moles of iodine reacted, the amount of formaldehyde can be determined.

$$HCHO + I_2 + 2NaOH \rightarrow 2NaI + H_2O + HCOOH$$

1 mole of HCHO ≡ 1 mole of I_2

30 g of HCHO ≡ 1 mole of I_2

15g of HCHO ≡ 1 equivalent of iodine

1000 ml of 1 N iodine solution ≡ 15 g of formaldehyde

1000 ml of N/10 iodine ≡ 1.5 g of formaldehyde

1 ml of N/10 iodine ≡ 0.0015 g of formaldehyde

Chemicals required

AR Potassium dichromate	= 1.2 g /250 ml water (about N/10)
Sodium thiosulphate	= 25 g /1 litre (about N/10)
Iodine solution	= 12.7 g/1 litre (about N/10)
Potassium iodide(10%)	= 10 g/100 ml
Starch solution(1%)	
Formalin solution	

Procedure

Standardisation of sodium thiosulphate solution:

A standard solution of N/10 $K_2Cr_2O_7$ is prepared by weighing about 1.2 g of analar crystals (eq. wt.= 49) accurately, dissolving in water and making up to 250 ml in a standard flask.

20 ml of standard $K_2Cr_2O_7$ solution is pipetted out into a clean conical flask. About 5 ml con.HCl is added, followed by 10 ml of 10% aq. KI solution. The liberated iodine is immediately titrated against sodium thiosulphate solution taken in the burette. When the solution acquires a straw yellow colour, 1 ml of freshly prepared starch added, and the titration is continued (in drops) with constant shaking. The end point is the colour change from blue to green (due to Cr^{3+}). The titration is repeated to get concordant values. The strength of thio is calculated.

$$Cr_2O_7^{2-} + 14\,H^+ + 6I^- \rightarrow 2Cr^{3+} + 7H_2O + 3I_2$$

$$2S_2O_3^{2-} + I_2 \rightarrow 2I^- + S_4O_6^{2-}$$

Standardisation of iodine solution

20 ml of iodine solution is pipetted out into a clean 250 ml conical flask. About 1 ml. starch is added. The solution is titrated against thio solution taken in the burette. The end point is the disappearance of blue colour (blue to colourless). The titration is repeated to get concordant values.

Estimation of formaldehyde

About 2 g of formalin are weighed accurately, transferred into a 250 ml standard flask and made up to the mark with distilled water. 20 ml of the solution are mixed with exactly 40 ml of N/10 iodine solution and 10% sodium hydroxide is added drop by. drop from a burette till the liquid becomes light yellow in colour. After standing for about 15 minutes, 5 ml of conc.hydrochloric acid are added and the unreacted iodine is titrated with thiosulphate using starch as indicator. The titration is repeated for concordancy.

Calcuiation:

$$HCHO + I_2 + 2NaOH \rightarrow 2NaI + H_2O + HCOOH$$

1 mole of HCHO ≡ 1 mole of I_2

1000 ml of 1 N iodine solution ≡ 15 g of formaldehyde

1000 ml of N/10 iodine ≡ 1.5 g of formaldehyde

1 ml of N/10 iodine ≡ 0.0015 g of formaldehyde

Let the strength of sodium thiosulphate solution = x N

Let the strength of iodine solution = y N

20 ml of iodine solution requires A ml of sodium thiosulphate

∴ 40 ml of iodine solution requires 2A ml of sodium thiosulphate

Let 20 ml of formaldehyde solution + 40 ml of iodine solution require B ml of sodium thiosulphate

∴ 20 ml of formaldehyde solution requires (2A – B) ml of sodium thiosulphate

$$\text{Strength of HCHO solution} = \frac{(2A-B) \times \text{Strength of thio}}{20}\ \text{N}$$

Amount of HCHO present in 100 ml of the made up solution

$$= \frac{(2A-B) \times \text{Strength of thio}}{20} \times \frac{15}{10}\ \text{g}$$

(13) ESTIMATION OF GLYCINE

Principle

Neutral amino acids like glycine which contain equal number of amino and carboxylic groups, cannot be estimated by directly titrating with standard sodium hydroxide solution due to the formation of zwitter ions in aqueous solution. Hence, the estimation is done by protecting NH_2 group with excess of neutralized formaldehyde solution (which reacts with the amino group of the amino acid, leaving the carboxylic group free) and then titrating with standard sodium hydroxide solution.

$$H_3N^+{-}CH_2{-}COO \; + OH \rightleftharpoons H_2N{-}CH_2{-}COO \; + H_2O$$

Zwitter ion

$$H_2N{-}CH_2{-}COO + HCHO \rightarrow HOCH_2{-}HN{-}CH_2{-}COO$$
$$\rightarrow CH_2{=}NCH_2COOH + H_2O$$

$$CH_2{=}N{-}CH_2COOH + Na^+OH \rightarrow CH_2{=}N{-}CH_2COONa + H_2O$$

Chemicals required

AR Oxalic acid	1.6 g/250 ml water
Sodium hydroxide	4 g/1 litre
Formalin solution (40%)	60 ml
Phenolphthalein	
Glycine	38 g/l litre

(20 ml volume range may be given to each student to be made up into 100 ml in a standard flask)

Reagents required

Preparation of standard oxalic acid solution: About 1.6 g of crystalline oxalic acid is weighed accurately, transferred into 250 ml standard flask and the solution is made up to the mark with distilled water.

Preparation of neutral formaldehyde solution: To the formaldehyde solution about 5 drops of phenolphthalein indicator is added followed the addition of dilute sodium hydroxide solution till the solution is just faintly pink.

Procedure

Standardisation of sodium hydroxide solution: 20 ml of the standard oxalic acid solution is pipetted out into a clean conical flask, a drop of phenolphthalein is added and the solution is titrated against sodium hydroxide taken in the burette. The end point is the appearance of pale permanent pink colour. From the strength of oxalic acid solution, the strength of sodium hydroxide solution is calculated.

Standardisation of glycine solution: The given glycine solution is made up into 100 ml in a standard flask. 20 ml of the solution is pipetted out into a clean conical flask. About 20 ml of neutral formalin solution is added slowly with constant shaking. 2 or 3 drops of phenolphthalein indicator is added and the solution is titrated against sodium hydroxide solution taken in the burette. The end point is the appearance of pale permanent pink colour.

Blank titration

A blank titration is performed with all solutions as described above except glycine solution.

Calculation

From the strength of oxalic acid solution, calculate the strength of sodium hydroxide solution.

Let the strength of sodium hydroxide solution = x N

Let the vol. of NaOH required by glycine solution = V_1 ml

Let the blank titre value without glycine solution = V_2 ml

Let the volume of glycine solution taken = 20 ml

∴ Volume of sodium hydroxide required by glycine solution = $(V_1 - V_2)$ ml

$$\therefore \text{ Strength of glycine solution} = \frac{N_{NaOH} \times V_{NaOH}}{V_{Glycine}}$$

$$= \frac{N_{NaOH} \times (V_1 - V_2)}{20}$$

Amount of glycine present in the given 100 ml of the solution

$$= \frac{N_{NaOH} \times (V_1 - V_2)}{20} \times \frac{75}{10} \text{ g}$$

(14) ESTIMATION OF NITROGEN

(Kjeldahl's method)

Principle

When an organic nitrogenous compound is digested with conc. sulphuric acid in presence of a suitable catalyst like potassium sulphate, copper sulphate etc., it gives ammonium sulphate. Ammonium sulphate when boiled with sodium or potassium hydroxide liberates ammonia gas. Ammonia evolved is directly passed into an excess of a standard hydrochloric or sulphuric acid solution. Some of the acid is neutralized and the remaining unreacted acid is titrated with a standard sodium hydroxide solution. Thus, the amount of acid neutralised by ammonia can be calculated which gives the amount of ammonia evolved and thus the percentage of nitrogen is calculated from the known weight of the nitrogenous organic compound.

$$H_2N\ CONH_2 + H_2SO_4 \xrightarrow{\text{Boil}} (NH_4)_2SO_4$$

$$(NH_4)_2SO_4 + 2NaOH \rightarrow Na_2SO_4 + 2NH_3 + 2H_2O$$

$$2NH_3 + H_2SO_4 \rightarrow (NH_4)_2SO_4$$

This method is good for the estimation of nitrogen in compounds like amion, azo, diazo, hydrazo, nitro, nitroso, nitrite compounds etc.

Chemicals required

K_2SO_4	12 g.
Conc.H_2SO_4	20 ml.
NaOH soln.	(50%)

Standard H_2SO_4 solution (N/5)
Standard NaOH solution (N/5)
Methyl orange (indicator)

Procedure

Standardisation of sodium hydroxide solution: About 8 g of crystalline oxalic acid is weighed accurately, transferred into 250 ml standard flask and the solution is made up to the mark with distilled water. 20 ml of the standard oxalic acid solution is pipetted out into a clean conical flask, few drops of phenolphthalein are added and the solution is titrated against sodium hydroxide taken in the burette. The end point is the appearance of pale permanent pink colour. From the strength of oxalic acid solution, the strength of sodium hydroxide solution is calculated.

Preparation of sulphuric acid solution

An approximately N/5 sulphuric acid is prepared by diluting 15 ml of concentrated sulphuric acid to 1 litre with distilled water and then standardized by titrating against a standard solution of sodium hydroxide soltuion as described above.

Estimation of Nitrogen

About 0.5g of the organic compound containing nitrogen is taken in a Kjeldhal's tube. About 15 ml of con. sulphuric acid is added and the content is digested by heating on a sand bath for 30 minutes in a fume cupboard. The content is cooled about 5 g of potassium sulphate is added as a catalyst to elevate the boiling point of sulphuric acid. The tube is then heated strongly for about 2 to 3 hours till all the solid matter is dissolved and a yellow colour solution is obtained.

The content in the Kjeldahl's tube is cooled and transferred into Kjeldahl's flask. A few glass beads are added to avoid bumping. About 50 ml of 20% sodium hydroxide solution is run to the flask until the solution is alkaline. The content is then boiled so that the ammonia gas liberated is passed through the condenser and then to 50 ml of standard sulphuric acid solution. After the reaction is over, the excess unreacted sulphuric acid is determined by back titrating with standard acid solution.

Calculation

Let the weight of nitrogen compound = W g

Let the normality of aq. sodium hydroxide = x N

Let the normality of sulphuric acid = y N

Let the unreacted sulphuric acid requires = V_1 ml of x N NaOH solution

Let 50 ml of standard sulphuric acid require = V_2 ml of x N NaOH solution

Reacted sulphuric acid $\equiv (V_2 - V_1)$ ml of x N NaOH

$$\text{Volume of sulphuric acid} \equiv \frac{(V_2 - V_1) \times N_{NaOH}}{N_{H_2SO_4}} = V \text{ ml}$$

1 equivalent of NaOH

$\equiv$ 1 equivalent of H_2SO_4
$\equiv$ 1 equivalent of NH_3
$\equiv$ 1 eqivalent of Nitrogen
$\equiv$ 14 g of Nitrogen

1000 ml of 1 N sulphuric acid $\equiv$ 14 g of Nitrogen

$$\therefore \text{V ml of y N sulphuric acid} \equiv \frac{14 \times V \times N_{H_2SO_4}}{1000} \text{ g of Nitrogen}$$

$\therefore$ Percentage of Nitrogen

$$\text{in the organic compound} \equiv \frac{14 \times V \times N_{H_2SO_4}}{1000} \times \frac{100}{W} \text{ g of Nitrogen}$$

(15) Estimation of Halogens

Principle

A known weight of the substance is heated with fuming nitric acid in a sealed tube, the organic material being oxidized to carbon dioxide and water, and halogen into halide ion. The latter is subsequently washed out of the tube, precipitated as silver halide and estimated as silver halide. The method is general for all organic compounds.

Procedure

A known mass(about 1g) of an organic compound is taken(ofcourse the amount depends on the % of X such that the weight of AgX should be about 0.2-0.3g) in a hard glass test tube and mixed with fuming nitric acid. The test tube is then sealed on the top and placed in a heater or in a oil bath and maintained a temperature of 260°C for 2 to 3 hours. The Carius tube is then cooled. Before opening the cold Carius tube, pay particular attention to the initial gentle warming of the capillary end of the tube to ensure that the liquid which has condensed in the capillary is driven back well into the wide portion of the tube, otherwise it will be lost with the escaping gases. The tube is then cut open and the contents are transferred into a beaker. The tube is then washed well with water free from halides ions several times and all the washings are collected in the same container. The solution is then diluted to 200 ml and heated. To the hot solution, about 30 ml of 5% alcoholic silver nitrate solution is added very slowly with constant stirring. The solution is then digested on a water bath for 30 minutes. The silver halide precipitate is filtered off at the pump, washed with water, dried in hot air oven and weighed. From the mass of silver halide precipitate the % of halogen in the organic compound is calculated gravimetrically.

Calculation

Chlorine

Let the mass of organic compound = W g
Let the mass of silver chloride = x g

143.5 g of AgCl contains 35.5 g of chlorine

$$\therefore \text{ x g of AgCl contains } \frac{35.5 \times Mass\ of\ AgCl}{143.5} \text{ g of chlorine}$$

$$\text{\% of Chlorine in the organic compound} = \frac{35.5 \times Mass\ of\ AgCl \times 100}{143.5 \times W}$$

Bromine

Let the mass of organic compound = W g
Let the mass of silver bromide = x g
188 g of AgBr contains 80 g of bromine

$$\therefore \text{ x g of AgBr contains } \frac{80 \times Mass\ of\ AgBr \times 100}{188} \text{ g of bromine}$$

$$\text{\% of bromine in the organic compound} = \frac{80 \times Mass\ of\ AgBr \times 100}{188 \times W}$$

Iodine

Let the mass of organic compound = W g
Let the mass of silver iodide = x g

235 g of AgI contains 127 g of iodine

$$\therefore \text{ x g of AgI contains } \frac{127 \times Mass\ of\ AgI}{235} \text{ g of iodine}$$

$$\text{\% of iodine in the organic compound} = \frac{127 \times Mass\ of\ AgI \times 100}{235 \times W}$$

(16) ESTIMATION OF CARBONYL GROUP (Determination of percentage purity of carbonyl compound)

Principle

All carbonyl compounds form condensation products readily with ammonia derivatives like hydroxyl amine, hydrazine, phenyl hydrazine or semicarbazide. Carbonyl compounds when treated with hydroxyl amine hydrochloride, form oximes, liberating hydrochloric acid, which can be titrated against standard solution of sodium hydroxide and thus the percentage purity of carbonyl compound may be determined.

$$R_2C{=}O + H_2N{-}OH.HCl \rightarrow R_2C{=}N{-}OH + H_2O + HCl$$

Chemicals required

Crystals of oxalic acid	8 g
Carbonyl compound	1 g
Hydroxylamine hydrochloride	3.5 g
Sodium hydroxide	20 g/1litre in water
Absolute ethyl alcohol	125 ml
Methyl alcohol	100 ml
Bromophenol blue indicator	

Procedure

Preparation of Hydroxylamine hydrochloride solution: 3.5 g of hydroxylamine hydrochloride is dissolved in 20 ml of distilled water and the solution is made up to 100 ml in standard flask with absolute ethyl alcohol.

Standardisation of sodium hydroxide solution: About 8 g of crystalline oxalic acid is weighed accurately, transferred into 250 ml standard flask and the solution is made up to the mark with distilled water. 20 ml of the standard oxalic acid solution is pipetted out into a clean conical flask, few drops of phenolphthalein are added and the solution is titrated against sodium hydroxide taken in the burette. The end point is the appearance of pale permanent pink colour. From the strength of oxalic acid solution, the strength of sodium hydroxide solution is calculated.

Preparation of bromophenol blue indicator: About 3-4 drops of a 5% alcoholic solution of bromophenol blue is mixed with 5 ml of pyridine and diluted to 250 ml with rectified spirit.

Estimation of carbonyl group: About 1 g of the given carbonyl compound is weighed accurately and transferred into 250 ml RB flask. About 50 ml of bromophenol blue is added followed by the addition of 25 ml of hydroxylamine hydrochloride. The RB flask is attached with a reflux condenser and refluexed for about 2 hours.

Blank titration

A blank titration is performed with all solutions as described above except carbonyl compound solution.

Both the test solution and blank solution are cooled and titrated against NaOH solution taken in the burette. The end point is the appearance of blue green colour.

Calculation

Let the weight of the carbonyl compound = W g

Let the titre value for the test solution = V_1 ml of NaOH

Let the titre value for the blank solution = V_2 ml of NaOH

∴ NaOH used for carbonyl compound = $(V_2 - V_1)$ ml

% Purity of carbonyl compound =

$$\frac{(V_2 - V_1) \times N_{NaOH} \times M.Wt.\ of\ carbonyl\ compound}{Weight\ of\ carbonyl\ compound \times 10}$$

(17) ESTIMATION OF HYDROXYL GROUP
(In alcohols and phenols by acetylation method)

Principle

An alcoholic or phenolic compound when treated with acetic anhydride dissolved in pyridine forms an acetyl derivative. Pyridine acts as a catalyst also. The excess of acetic anhydride is hydrolysed to acetic acid with water. The acetic acid so formed is estimated by titration against a standard solution of sodium hydroxide using phenolphthalein s indicator.

$$R(OH)_n + n(CH_3CO)_2O \rightarrow R(OCOCH_3)_n + nCH_3COOH$$

A blank experiment is also performed using the same amount of acetic anhydride in pyridine (without using the alcoholic compound). The difference in the volumes of sodium hydroxide used in the two experiments is equivalent to the acetic acid used in acetylation.

Chemicals required

Alcoholic or phenolic compound	1 g.
Acetic anhydride	10 ml
Pure pyridine (anhydrous)	40 ml
Sodium hydroxide solution (N/10)	4 g/1 litre water
Phenolphthalein	
Crystalline oxalic acid	1.6 g

Procedure

Preparation of acetylating mixture: 10 ml of acetic anhydride is mixed with 40 ml pyridine thoroughly in a beaker and quantitatively transferred in to a burette attached with an anhydrous calcium chloride tube at the top.

Preparation of standard oxalic acid solution: About 1.6 g of crystalline oxalic acid is weighed accurately, transferred into 250 ml standard flask and the solution is made up to the mark with distilled water.

Standardisation of sodium hydroxide solution: 20 ml of the standard oxalic acid solution is pipetted out into a clean conical flask, a drop of phenolphthalein is added and the solution is titrated against sodium hydroxide taken in the burette. The end point is the appearance of pale permanent pink colour. From the strength of oxalic acid solution, the strength of sodium hydroxide solution is calculated.

Acetylation: About 0.5 g of the given hydroxy compound is weighed accurately and transferred into a conical flask fitted with a reflux water condenser. Exactly 20 ml of the acetylating mixture is added slowly with constant stirring from the burette. The content is then refluxed on a boiling water bath for 15 minutes. The content is transferred slowly with shaking into 100 ml standard flask and made up to the mark with distilled water.

Estimation of hydroxy group: 20 ml of the reaction mixture is pipetted out into a clean conical flask. 2 to 3 drops of phenolphthalein indicator is added. The reaction mixture is then titrated against sodium hydroxide taken in the burette. The end point the appearance of pale permanent pink colour(During the titration a precipitate of acetyl compound may be obtained but it has no effect on titration).

Blank titration: Exactly 20 ml of the acetylating mixture is taken in a conical flask and refluxed on a boiling water bath for 15 minutes. The content is transferred slowly with shaking into 100 ml standard flask and made up to the mark with distilled water. 20 ml of this solution is pipetted out into a clean conical flask. 2 to 3 drops of phenolphthalein indicator is added. The reaction mixture is then titrated against sodium hydroxide taken in the burette. The end point the appearance of pale permanent pink colour A blank titration is performed with all solutions as described above except hydroxy compound solution.

Calculation

Weight of the alcoholic compound = W g

Let the normality of NaOH solution = x N

20 ml of diluted reaction mixture ≡ V_1 ml of x N NaOH solution.

20 ml of diluted blank solution ≡ V_2 ml of x N NaOH solution

Volume of x N NaOH solution equivalent to the actual acetic acid consumed in the acetylation = $(V_2 - V_1)$ ml. of x N NaOH solution

1000 ml of N NaOH ≡ 1 gram mole of NaOH
≡ 1 gram mole of CH_3COOH
≡ 1 hydroxyl group

$(V_2 - V_1)$ ml. of x N NaOH solution $\frac{1}{1000} \times (V_2 - V_1) \times N_{NaOH}$

W g of the alcoholic or phenolic compound contains

$$= \frac{1}{1000} \times (V_2 - V_1) \times N\ NaOH \text{ hydroxyl groups.}$$

∴ 100 g of the alcoholic or phenolic compound contains

$$= \frac{1}{1000} \times (V_2 - V_1) \times N_{NaOH} \times \frac{100}{W} \text{ hydroxyl groups}$$

Note: The heating of the reaction mixtures (with or without alcoholic or phenolic compound) may be performed simultaneously and separately in two different flasks.

(18) DETERMINATION OF EQUIVALENT WEIGHT OF AN ACID

Principle

The equivalent weight of an acid is the ratio between the molecular mass and its basicity. The basicity is the number of replaceable hydrogen atoms, present in an acid. The equivalent weight of a water soluble acid can be determined by titrating it against a standard solution of an alkali using phenolphthalein as indicator.

$$R(COOH)_n + nNaOH \rightarrow R(COONa)_n + nH_2O$$

Chemicals required

Oxalic acid Crystals	3.2 g
Sodium hydroxide solution(N/5)	20g/1 litre water
Carboxylic acid(eq.wt. to be determined)	3 g
Phenolphthalein	

Procedure

Standardisation of sodium hydroxide solution: A standard solution of N/5 oxalic acid is prepared by weighing about 3.2 g of analar crystals (eq. wt.= 63) accurately, dissolving in water and making up to 100 ml in a standard flask.

20 ml of standard oxalic acid solution is pipetted out into a clean conical flask. About 2 drops of phenolphthalein is added. The solution is then titrated against sodium hydroxide solution taken in the burette. The end point is the appearance of pale permanent pink colour. The titration is repeated for concordancy. From the strength of oxalic acid solution, the strength of sodium hydroxide is dertermined.

Standardisation of carboxylic acid: About 2 to 2.5 g of the given water soluble carboxylic acid whose equivalent mass to be determined is weighed accurately, dissolved in water and the solution is made up into 100 ml in a standard flask.

20 ml of the made up carboxylic acid solution is pipetted out into a clean conical flask. About 2 drops of phenolphthalein is added. The solution is then titrated against sodium hydroxide solution taken in the burette. The end point is the appearance of pale permanent pink colour. The titration is repeated for concordancy. From the strength of sodium hydroxide solution, the strength of carboxylic acid is dertermined.

Calculation

Let the strength of sodium hydroxide = x N

Let the weight of carboxylic acid taken = W g

20 ml of carboxylic acid solution requires V ml of x N NaOH solution

∴ 100 ml of carboxylic acid solution requires 5V ml of x N NaOH solution

1 equivalent of a base ≡ 1 equivalent of an acid

5V ml of x N NaOH solution ≡ W g of Carboxylic acid

∴ 1000 ml of 1N NaOH solution ≡ $W \times \frac{1000}{5V} \times \frac{1}{x}$ g of Carboxylic acid

Therefore, the equivalent weight of the acid = $W \times \frac{1000}{5V} \times \frac{1}{x}$

(19) Reichert-Meissel Value (R.M. Value) of an Oil (or) Fat

Principle

The number of milliliters of decinormal potassium hydroxide solution required to neutralise the volatile acids obtained by distillation of 5 g of an oil fat is called Reichert-Meissel value shortly called R-M value of an oil or fat. This gives an estimate of the lower steam volatile fatty acids present in an oil or a fat. It is important for testing the purity of butter or ghee. The R.M. value of butter or animal ghee (20-30) is lowered on adulteration.

Chemicals required:

Oil, ghee or butter	5 g
Alcoholic sodium hydroxide (N/2)	200 ml(6 g of KOH dissolved in 200 ml of rectified spirit)
N/2 Hydrochloric acid	100 ml(10 ml of con.HCl in 200 ml)
Phenolphthalein	
AR sodium carbonate crystals	7 g

Procedure

Standardisation of HCl

About 7 g of crystals of anhydrous sodium carbonate is weighed accurately, dissolved in water and the solution is made upto 250 ml in a standard flask. 20 ml of this solution is pipetted out into a clean conical flask, 2 drops of methyl orange indicator are added and titrated against hydrochloric acid taken. The end point is the appearance of light pink colour. The titration is repeated for concordancy and the strength of hydrochloric acid is calculated.

Determination of R-M Value

About 2 g of the given oil is weighed accurately and taken in a 250 ml RB flask. Exactly 25 ml of alcoholic KOH solution is added to it. The solution is refluxed on a water bath for 30 minutes. A blank is run simultaneously with the same quantity of alcoholic potash but without oil. Both the flasks are cooled and titrated against standard HCl solution using 1 ml phenolphthalein indicator. The end point the disappearance of pink colour.

Calculation:

Calculate the strength of HCl from the first titration.

Let the weight of oil taken = W g

Let the volume of HCl required by test solution = V_1 ml

Let the volume of HCl required by blank solution = V_2 ml

∴ Volume of alcoholic KOH required by the oil ≡ $(V_2 - V_1)$ ml of HCl

$$\text{R-M value of the given oil} = \frac{(V_2 - V_1)}{W} \times 5 \times \frac{0.1}{Str.\ of\ HCl}$$

PART E
EXTRACTION OF NATURAL PRODUCTS

(01) ISOLATION OF CAFFEINE FROM TEA

Requirements

Chemicals	*Apparatus*
Tea dust or fluff 250 g	Round-bottom flask 1 litre
Lead acetate 20g/100 ml water	Wire gauze, Beaker 250 ml
Litharge 20 g	Cotton cloth or tea filter

Procedure

Basic lead acetate solution is prepared by boiling lead acetate solution with litharge with continuous stirring. The solution is then filtered and heated on boiling water bath.

To the tea dust, about 500 ml of water is added in a round bottom flask and the contents boiled for half an hour on a wire gauze. The tea content is filtered on a tea filter or a cotton cloth. To the filtrate basic lead acetate solution is added slowly with constant stirring till no further precipitation of tannin takes place.

The mixture is heated on a wire gauze to boil, filtered and the precipitate is washed with hot water. The excess of lead is removed by adding dilute sulphuric acid drop by drop to the mixture and allowing the precipitate to settle. The clear supernatant liquid is decanted and concentrated by boiling.

The filtrate after cooling is extracted with 20 ml of chloroform atleast three times and the chloroform is distilled from the combined extract. The residue is boiled with water and filtered. The filtrate is set aside for crystallization. The caffeine crystallizes out as white crystals .

The crude caffeine is further purified by vacuum sublimation.

(Expected yield: 3 g; Melting point of anhydrous caffeine: 237°C)

(02) Isolation of Lactose from Milk

Requirements

Chemicals		*Apparatus*
Milk	500 ml	Beaker 1 litre
Dilute acetic acid	100 ml	Centrifuging tubes and machine
Calcium carbonate	2 g	Wire gauze, Funnel, glass rod
Rectified spirit	50 ml	Watch glass

Procedure

About 500 ml of milk is diluted with water and centrifuged to remove the fat. The fat appears on the top of the centrifuging tube is removed. The clear thin milk obtained after centrifugation is called skim milk. To the clear skim milk about 50 ml of dilute solution of acetic acid is added slowly with constant stirring. A white solid separates out called casein. The precipitated casein is filtered off, and the filtrate is neutralized with dilute aqueous sodium hydroxide solution. The solution is made very slightly acidic by adding 2 or 3 drops of acetic acid to the neutralized solution. The solution is boiled on a wire gauze with 2 g of calcium carbonate solid. The mixture is stirred well and filtered hot. The solution is concentrated by boiling and converted into one third volume. About 3 volumes of rectified spirit is added and the solution is filtered. White crystals of lactose separates out. The white solid is filtered off at the pump, washed with alcohol and dried in air.

(Expected yield: 2 g)

(03) Isolation of Citric Acid from Lemon

Requirements

Chemicals		*Apparatus*
Lemon juice	100ml	Beaker 500 ml
Calcium chloride(10%)	2 ml	Watch glass
Dilute ammonia	15 ml	Glass rod
Dilute sulphuric acid	10 ml	Funnel

Procedure

About 100 ml of either lemon juice or pineapple juice is taken and diluted with double the amount of water. The juice is neutralized by adding dilute ammonia slowly with constant stirring. The citric acid present in the juice is now converted into ammonium citrate. To the solution of ammonium citrate, about 2 ml of a 10% aqueous solution of calcium chloride is added. The resulting solution is heated to boiling and set aside for 30 minutes. A white precipitate of tricalcium citrate is formed. The precipitate is washed well with water and the washing is decanted. From the tricalcium citrate, citric acid can be obtained by treating with dilute sulphuric acid. The white ppt of calcium sulphate formed is filtered and the filtrate is concentrated where crystals of citic acid appears.